Columbia University

Contributions to Education

Teachers College Series

No. 925

AMS PRESS
NEW YORK

ONE-STEP PROBLEM PATTERNS AND THEIR RELATION TO PROBLEM SOLVING IN ARITHMETIC

By ETHEL SUTHERLAND, Ph.D.

TEACHERS COLLEGE, COLUMBIA UNIVERSITY
CONTRIBUTIONS TO EDUCATION, NO. 925

Published with the Approval of
Professor C. B. Upton, Sponsor

BUREAU OF PUBLICATIONS

TEACHERS COLLEGE, COLUMBIA UNIVERSITY

NEW YORK, 1947

Library of Congress Cataloging in Publication Data

Sutherland, Mary Ethel, 1889-
One-step problem patterns and their relation to problem solving in arithemtic.

Reprint of the 1947 ed., issued in series: Teachers College, Columbia University. Contributions to education, no. 925.
Originally presented as the author's thesis, Columbia.
Bibliography: p.
1. Arithmetic--Study and teaching. 2. Arithmetic--Problems, exercises, etc. I. Title. II. Series: Columbia University. Teachers College. Contributions to education, no. 925.
QA135.5.S93 1972 372.7'2'044 76-177734
ISBN 0-404-55925-5

From the edition of 1947, New York
First AMS edition published in 1972
Manufactured in the United States

AMS PRESS, INC.
NEW YORK, N. Y. 10003

ACKNOWLEDGMENTS

THE AUTHOR wishes to express thanks to her sponsor, Professor Clifford B. Upton, for his inspiration, encouragement, and generous help throughout the progress of this study.

The author also wishes to express gratitude to Professor Robert L. Thorndike for his friendly counsel and constructive suggestions in the planning of this study. In addition, the author expresses appreciation to Professor William D. Reeve and Professor John R. Clark for helpful suggestions; and to Miss Margaret Uhlinger for invaluable aid in collecting and tabulating the data.

E. S.

CONTENTS

ONE-STEP PROBLEM PATTERNS AND THEIR RELATION TO PROBLEM SOLVING IN ARITHMETIC

CHAPTER I

THE PROBLEM

INSTRUCTION IN PROBLEM SOLVING in arithmetic has always presented difficulties both to teachers and to pupils and the results have generally been unsatisfactory. Newcomb states that "the attention of every teacher of mathematics has at some time or other been directed to the fact that pupils have unusual difficulties in problem-solving in arithmetic, and that they achieve considerably more unsatisfactory results in this connection than with computation. The truth of this fact has been thoroughly substantiated by the results obtained in numerous scientific tests."[1] Stone agrees that "it is the experience of teachers that to teach a pupil how to solve a problem is a much more difficult task than to train for skill in computation";[2] Durell adds that "teachers generally admit that they have more difficulty in teaching problems than abstract processes and the results are less satisfactory."[3] Welch and Shibles confirm these statements by asserting that "no phase of the curriculum offers greater challenge to teachers and supervisors than does problem work in arithmetic."[4]

Research concerning the teaching of arithmetic has been far more concerned with the computational side of this subject than with problem solving. As Lutes remarks, "The significant fact concerning the scientific literature of problem solving is its paucity; little experimentation is on record that has deliberately attempted to improve the ability of school pupils

[1] R. S. Newcomb, *Modern Methods of Teaching Arithmetic,* p. 270. 1926.

[2] J. C. Stone, *The Teaching of Arithmetic,* p. 174. 1918.

[3] T. J. Durell, "Problem Solving in Arithmetic," *1941 Yearbook, Association of Mathematics Teachers of New Jersey,* p. 34. 1941.

[4] G. L. Welch and M. R. Shibles, "A Fifth Grade Informal Problem Unit in Arithmetic," *Education,* 61 : 458, 1941.

to solve verbal problems."[5] Hanna[6] made a careful analysis of 482 investigations in arithmetic that were reported by Buswell and Judd,[7] including those investigations reported later by Buswell,[8] and found that only 6 per cent of these studies dealt mainly with problem solving; Hanna's analysis covered investigations up to 1928. Since that time the relative attention given to problem solving has changed only slightly for the better.

The earlier studies in problem solving were concerned largely with the elimination of obsolete problems and the attempt to make problems more practical. Later studies have been concerned with the specific causes of success and failure in problem solving; the techniques for increasing ability to solve problems; the relation of reading ability to problem-solving ability; the relation between ability in computation and ability in problem solving; the influence of language and vocabulary upon the ability to solve problems; and the influence of verbal cues upon problem solving.

A review of all the literature on problem solving up to 1925 is given by Buswell and Judd in their comprehensive *Summary of Educational Investigations Relating to Arithmetic.*[9] Studies since 1925 are reported annually by Buswell in the *Elementary School Journal* under the caption "Summary of Arithmetic Investigations."

To date no study in problem solving has been made which attempts to classify the verbal problems in elementary arithmetic textbooks in terms of one-step patterns and to show the bearing of such patterns on the teaching of problem solving. That is the purpose of the present study.

Before describing this study it is necessary to define problems in terms of the number of steps that are required to solve

[5] O. S. Lutes, *An Evaluation of Three Techniques for Improving Ability to Solve Arithmetic Problems,* p. 10. 1926.

[6] P. R. Hanna, *Arithmetic Problem Solving,* p. 43. 1929.

[7] G. T. Buswell and C. H. Judd, *Summary of Educational Investigations Relating to Arithmetic.* 1925.

[8] G. T. Buswell, "Summary of Arithmetic Investigations," *Elementary School Journal,* Vols. 28–30, 1926, 1927, 1928.

[9] Buswell and Judd, *loc. cit.*

them. A one-step problem, as that term is used in this study, refers to a problem whose answer is obtained by the single performance of one of the four fundamental operations of addition, subtraction, multiplication, and division. The following is an illustration of a one-step problem: "John sold 15 papers this morning and 20 this afternoon. How many papers in all did he sell today?" A single addition is sufficient to solve this problem. A two-step problem is one in which it is necessary to perform two successive operations in order to obtain the answer. The following is a two-step problem: "Mary has 25¢. If she buys a pencil for 5¢ and an eraser for 2¢, how much money will she have left?" In this problem, there is first an addition and then a subtraction. It is evident that the solution of this two-step problem breaks down into the solution of two one-step problems. A three-step problem, in like manner, requires three successive operations in order to obtain the answer. Each of the three steps in such a problem is a one-step problem. That is, to solve a three-step problem one has successively to solve three one-step problems. It is thus seen that a pupil must know how to solve one-step problems before he can solve problems of two or more steps. For convenience in this study problems of two or more steps are called multiple-step problems.

It will be helpful at this point to give a few illustrations of one-step patterns. The following are examples of such patterns.

(1) *Tom had 40¢. He spent 25¢. How many cents did he have left?*

This problem is of the following pattern: A given amount has been decreased. How much is left? The amount left is found by subtraction. In problems of this pattern, the notion of a *decrease* may be suggested by a wide variety of words, such as *spend, give, lose, fly away,* etc. The phrases *how many are left* and *how many remain* are also characteristic of this pattern. It is evident that there are a great many one-step problems that conform to this pattern.

(2) *Joe has 30¢. Mary has 25¢. How many more cents has Joe than Mary?*

This problem follows the pattern: Two unequal amounts are given which are to be compared by finding *how much more* one amount is than the other. The comparison may also be made by finding *how much less* one amount is than the other. In problems of this pattern, the comparison is made by subtraction. The expressions *how many more* and *how many less* are characteristic of this pattern.

(3) *Uncle Joe had 45 stamps which he divided equally among 3 boys. How many stamps did each boy get?*

This problem follows the pattern: A given amount is to be separated into equal parts. What is the size of each part? The size of each part is found by division. In this pattern the separation into equal parts is suggested by such phrases as *divide equally among* and *what is each boy's share.* This pattern represents the sharing concept.

(4) *The Empire State Building is 1250 ft. high. The Franklin Bank is 208 ft. high. The Empire State Building is about how many times as high as the Franklin Bank?*

This problem follows the pattern: Two unequal amounts are given which are to be compared by finding how many times as large one amount is as the other. In this case the comparison is made by division. The expressions *how many times as large, how many times as many, how many times as high,* etc., are characteristic of this pattern.

The four problems given above illustrate four fundamental one-step patterns, two of them being related to *subtraction* and two to *division.* There are still other patterns that are associated with each of the four fundamental operations of arithmetic. In this study thirty-eight patterns in all have been found. These thirty-eight patterns cover all four fundamental operations. Some of these patterns are closely identified with certain life activities.

In each of these thirty-eight patterns the basic ideas or concepts represented are of outstanding importance. The thought element dominant in certain patterns is often expressed in technical language with which the pupil must become familiar if he is to understand the patterns. This technical language, however, is not the dominating feature of the pattern. It is the thought or concept behind the pattern that is important. In this study it must be made clear that the technical phraseology peculiar to any pattern is not to be used mechanically as a verbal cue or hint that suggests to the pupil the arithmetic operation to be performed. Problems representing each one-step pattern must have reality and meaning for the pupil; such problems must also be within his understanding and experience. If a problem representing a new pattern is unrelated to a pupil's activities and experiences, these activities must, at an appropriate time, be enlarged and enriched until the pattern comes to have full meaning.

Statement of the Problem

The problem with which this study is concerned will now be stated in greater detail. It consists of three tasks as follows:

I. *To determine how many different one-step patterns there are in connection with each of the four operations, addition, subtraction, multiplication, and division.*

This was done by examining all the verbal problems in a group of modern basal textbooks in arithmetic covering the work of Grades 3 to 6 inclusive. Problems of two or more steps were included in this examination in order to determine the one-step elements. In this study it is assumed that the authors of any given series of arithmetics, covering Grades 3 to 6, have provided in their series a sufficient amount and variety of problem material to teach the pupil to solve problems in arithmetic of a level of difficulty suitable for these grades.

II. *To determine for each series of texts examined the fre-*

quency with which the one-step patterns occur in the problems of each grade.

This made it possible to determine whether the authors of arithmetics tend to have any agreement or uniformity in their problem material with respect to the fundamental patterns that are appropriate to each grade level. It also made it possible to find out whether important patterns receive relatively little or no attention.

III. *To discuss the possible bearing of the above findings upon the teaching of problem solving in arithmetic.*

On the basis of the problem patterns found in this study it was possible to offer suggestions concerning those types of quantitative thinking that are appropriate at certain grade levels.

CHAPTER II

PROCEDURE

IN ORDER TO DETERMINE the number of one-step patterns in connection with each of the four fundamental operations, the verbal problems of four modern, widely used, basal textbooks in arithmetic, covering Grades 3 to 6 inclusive, were examined. Each of these series of arithmetics consisted of four separate books, one for each grade. Hence, sixteen different books in all were analyzed. In deciding upon the series of arithmetics to be used in this analysis, it became necessary to choose those that followed a uniform course of study in arithmetic of a modern type. Fortunately this condition was easily satisfied by choosing the state course of study in arithmetic of one of the most prominent states. This course of study was issued in 1937 and represents the newer type of curriculum which has been widely accepted in elementary schools throughout the United States. Three of the series of arithmetics selected for analysis were editions of well-known textbooks prepared especially to meet this course of study. The fourth series, while not especially prepared for that purpose, also met the requirements of the syllabus.

Since verbal problems often relate to the abstract processes that are being taught in a given grade, it was essential, of course, that the textbooks follow a common course of study, otherwise the comparisons of patterns made later in this study would not have been trustworthy. While the textbooks were chosen primarily with reference to the course of study, all these books possessed several other desirable features which tend to give assurance that the problem material in these books is representative of the best that is available. These other features are as follows:

(1) The authors of all four series are arithmetic specialists

of national reputation. These authors are familiar with recent tendencies and improvements with respect to problem solving in arithmetic.

(2) The problem material in all these books is modern, interesting, and related to everyday experiences of children.

(3) Each of the series of books is a revision of, or a successor to, a previous series of books by the same authors. The fact that any textbook passes into a second edition is generally one evidence of its usability.

(4) The textbooks in question are among the most widely used textbooks in arithmetic in the United States.

In examining the verbal problems in these four series of arithmetics all the problems found in each book were studied with the exception of a very few which will be mentioned later. The findings of this study, therefore, are not based on random samples but on the complete offering in verbal problems in each of these series. All together the four series, covering Grades 3 to 6 inclusive, had a total of about 15,000 verbal problems, all of which were analyzed in gathering the data in this study. The analysis covered one-step problems and multiple-step problems. Each one-step problem was at once classified in terms of the one-step patterns discussed in this study. Each multiple-step problem was broken down into its one-step elements, which in turn were classified in terms of one-step patterns.

At the beginning of this study an exploratory analysis was made which disclosed the fact that some limitations would have to be placed upon the study in order to keep it within reasonable limits. With the approval of the sponsor and other advisers, it was decided to omit the following items:

(1) All problems involving the mere counting of change.

(2) All mensuration problems requiring the determination of perimeter, area, or volume.

(3) All problems in denominate numbers expressed in, and requiring an operation on, two or more denominations. Thus problems requiring the addition of 6 ft. 3 in. to

8 ft. 2 in. or the multiplication of 8 ft. 5 in. by 3 were omitted.

(4) All problems involved in making graphs and scale drawings.

(5) All problems involving Case II of percentage. Problems involving Case I of percentage are included.

(6) All incomplete problems, such as those in which data or a final question have to be supplied before the problem can be solved.

(7) All multiple-step problems involving four or more steps. This limitation on multiple-step problems was made after an initial exploration had shown that no new patterns were found in problems of four or more steps that had not appeared in problems of two and three steps.

The omissions above represent only a small fraction of the total problem material in each series of books. This is evident from the fact that the verbal problems which were finally included in this study yielded a total of 13,498 repetitions of the 38 one-step patterns found as a result of the investigation.

To make the results of this study as dependable as possible, the selection of the pattern or patterns represented in each problem was checked, in all cases of doubt, by a second person and sometimes by a third. In the case of one-step problems, there were relatively few cases where the identity of the pattern was in question; in such cases a second or third person independently classified the problem. The analysis of all two-step and three-step problems in the entire study was checked independently by a second or third person. If another investigator should classify the problems in the four series of books in terms of the patterns found in this study, it seems unlikely that the results would differ materially from those reported herein.

Examples will now be given of the more important questions that had to be answered with respect to the classification of certain types of problems. In a few instances, a problem might be classified as a one- or two-step problem, a two- or three-step problem, or a three- or four-step problem, depending upon the way it is worked. For example, a child might work the following problem in either of two ways: "Helen bought 3 yards of lace and 2 yards of ribbon, both at 40¢ a yard. How much did she pay for them?" The problem could be solved by multiplying 40¢ by 3 and 40¢ by 2 and then adding $1.20 and $.80, or by first adding 3 and 2 and then multiplying 40¢ by 5. The first solution involves three steps (2 multiplications and 1 addition); the second solution involves two steps (1 addition and 1 multiplication). Both solutions involve the same patterns. The second solution is the more economical solution and the more likely. All such problems, of which there were about fifty, were counted as two-step problems.

The following problem might be solved in either of two ways: "Jack and Tom and their mother and father went to Troy on the train. Jack, Tom, and their mother had the 75-cent lunch and their father the 95-cent lunch. How much did Jack's father pay for all the lunches?" This problem could be solved by adding 75¢ + 75¢ + 75¢ + 95¢ or by multiplying 75¢ by 3 and then adding $2.25 and $.95. The first solution involves one step, which is addition; the second solution involves two steps (1 multiplication and 1 addition). Problems of this type, about six in number, were counted as two-step problems. The following problem is an illustration of a three-step or a four-step problem, depending upon the way it is worked: "Lena is saving to buy her mother an electric iron costing $3.95 for Christmas, which is 13 weeks off. If she saves 10¢ a week on moving pictures and 20¢ a week on candy how much more will she need to buy the iron?" The problem could be solved by multiplying 10¢ by 13 to find the saving on pictures, 20¢ by 13 to find the saving on candy, by adding $1.30 and $2.60 to find the amount saved, by subtracting $3.90 from $3.95 to find how much more money is needed. This solution involves four steps. The more natural way to solve it would be to add 10¢ and 20¢, to multiply 30¢ by 13, and to subtract $3.90 from $3.95. The first solution involves four steps (2 multiplications, 1 addition, and 1 subtraction); the second solution involves three steps (1 multiplication, 1 addition, and 1 subtraction). Both solutions make use of the same operations, but the second is the more economical and the more probable. All problems such as this were counted as three-step problems. The number of such problems was very small, probably not exceeding ten.

There are a few types of problems which may be solved two different ways, but which yield the same number of steps by either method and sometimes the same operations. The following problems are illustrations:

(1) How much will 18 oranges cost if you can buy 3 oranges for 10¢?
(2) At a sale I bought a sweater for 1/4 off. If the original price of the sweater was $4.00, what did I pay for it?
(3) If a store sells a dress at a discount of 35%, how much will the dress cost if the original price was $16?
(4) Ann got $5.00 for a present. She bought a pair of roller skates for $2.75 and a book for $1.19. How much money did she have left?
(5) David bought 4 dozen small plants at 25¢ apiece for his garden. Find his bill for these plants.

The first problem might be solved by either the ratio method or the method of unitary analysis. By the ratio method the solution would be something like this: "18 oranges is 6 times as many as 3 oranges, so 18 oranges will cost 6 times as much as 3 oranges, or 6 × 10¢." By the method of unitary analysis the solution would be something like this: "If 3 oranges cost 10¢, 1 orange will cost 1/3 of 10¢, or 3 1/3¢. 18 oranges will cost 18 × 3 1/3¢, or 60¢." Both solutions involve division in the first or hidden step and multiplication in the final step. In deciding which of these two methods to select, the writer was governed by the model solutions used in each of the four series of textbooks. If a textbook used the method of unitary analysis in its model solutions of these problems, that method was kept in mind when the patterns were selected. If a textbook used the ratio method in its model solutions of these problems, those patterns were selected that would harmonize with the ratio method.

The second and third problems are very similar. In problem 2, the term "1/4 off" is used, whereas in problem 3, the term "discount of 35%" is used. These problems might be worked either of two ways. One could subtract first and then multiply, or one could multiply first and then subtract. For example, in problem 1, one could take 1/4 off, leaving 3/4, and then find 3/4 of $4, or one could take 1/4 of $4, which is $1, and subtract 1 from 4; in problem 2, one could subtract 35% from 100% and then multiply $16 by 65%, or one could find 35% of $16 and then subtract $5.60 from $16. The same operations are used in both solutions, but their order of appearance is reversed. To multiply first and then subtract seems the more natural solution for a pupil learning to solve these problems. All problems such as these were therefore credited to the second solution.

The usual way of solving problem 4 would be to add $2.75 and $1.19 and subtract their sum from $5.00. Some children might solve this problem by subtracting $2.75 from $5.00 and then subtracting $1.19 from the amount left. There is no way of telling which method a child would use, but the first seems more natural. Such problems were therefore credited to the method which first used addition and then subtraction. The number of such problems was very small.

In problem 5 one cannot be sure whether a child will first find how many plants there are in 4 dozen and then multiply 25¢ by 48, or whether he will find the cost of 1 dozen first and then multiply $3 by 4 to find the cost of 4 dozen. In either solution, he would use the operation of multiplication twice. Such problems as these were arbitrarily credited to the first method. The number of such problems was very small.

This chapter has described in some detail the procedure that was followed in analyzing the textbooks, the types of problems that were included or omitted, the difficulties that were encountered in determining the number of steps in certain types of problems, and the method that was selected when a problem could be solved in more than one way. In the following chapters, the patterns associated with each of the four fundamental operations are described in detail and their treatment in each of the series of textbooks is discussed.

CHAPTER III

SUBTRACTION PATTERNS

A CRITICAL EXAMINATION of all the subtraction problems found in the four series of textbooks analyzed in this study shows that these problems can be broken down into ten one-step patterns. Some of these patterns are related to each other in their underlying thought and can be grouped together, thus making four major groups of patterns—those patterns which involve (I) the remainder or "how-much-left" idea, (II) the "how-much-more" or building-up idea, (III) the comparison or difference idea, and (IV) the "separation-into-parts" idea. The differences among these four groups of patterns are primarily differences in their underlying thought, whereas the differences among the patterns within a given group are sometimes concerned with the technical phraseology peculiar to each pattern. The four groups of patterns and the patterns within each group are illustrated below.

I. THE PATTERNS THAT INVOLVE THE REMAINDER OR "HOW-MUCH-LEFT" IDEA

S1. A *given amount has been decreased.* Find *how much* is *left* or *remains.*

The notion of a decrease is suggested by a wide variety of words, such as *spend, sell, give, take, lose, fly away, run away, break, eat,* etc. *How many are left* or *how many remain* are questions characteristic of this pattern. The following problems are used to illustrate this pattern:

(1) Jack has saved 69¢. He buys a ball that costs 55¢. *How much* money has he *left*?

(2) Jim had 83 marbles. He gave 35 marbles to Bob. *How many* marbles did he have *left*?

S2. A *given amount* of money *has been decreased* by a purchase. Find the *amount of change received.*
How much change was received is the question characteristic of this pattern.

(1) At the store Joe paid $.73 for a book. He gave the clerk $1.00. *How much change* did he get?

(2) Mother bought a piece of meat for 35¢. *How much change* should she receive from a 50-cent piece?

S3. A *given amount has been decreased.* The *amount left* or the *change received* is known. Find the *amount taken away, spent, sold, given away,* etc.
This is the inverse of Patterns S1 and S2. In this pattern a given amount has been decreased but one finds the *amount taken away* instead of the *amount left.* Since the *amount left* is given, one works backward, as it were, to find the *amount spent, given away, eaten, sold,* etc.

(1) When Helen bought her book she gave the clerk a dollar and he gave her 25¢ change. How much did Helen's book cost?

(2) Yesterday Ruth's mother bought a dozen oranges. Today all but 4 of the oranges have been eaten. How many oranges have been eaten?

The second group, which is suggestive of addition, contains only one pattern as follows:

II. THE PATTERN THAT INVOLVES THE "HOW-MUCH-MORE" OR BUILDING-UP IDEA

S4. A *given amount* is on hand and a *given larger amount* is desired. Find *how much more* is needed to equal the desired larger amount.

(1) Betty wants to color 12 eggs for Easter. She has colored 4 eggs. How many does she *still need* to color?

(2) John wants to buy an airplane for 75 cents. He has only 69 cents in his bank. *How much more* money does he need to buy the airplane?

Pattern S4 is closely related to the additive method of subtraction. This is the only subtraction pattern in which the underlying thought relates to the building-up process.

The third group of patterns is concerned with the various ways of comparing two numbers by subtraction. These patterns are illustrated below.

III. THE PATTERNS THAT INVOLVE THE COMPARISON OR DIFFERENCE IDEA

S5. *Two unequal amounts are given.* They are to be compared by finding *how much more* or *how much less* one is *than* the other.

Characteristic expressions are *how many more than, how much less than.*

(1) For the school party Mary made 14 boxes of candy and Jane made 8 boxes of candy. *How many more* boxes of candy did Mary make *than* Jane?

(2) Betty weighs 56 pounds. Dotty weighs 63 pounds. Betty weighs ..... pounds *less than* Dotty.

S6. *Two unequal amounts are given.* They are to be compared by finding how much *larger, longer, taller, older,* etc., one is *than* the other.

Characteristic expressions are *how much older, how much taller, how much heavier,* etc.

(1) Harry is 10 years old. His father is 45 years old. *How many years older than* Harry is his father?

(2) This morning our thermometer read 65°. This afternoon it read 72°. *How many degrees warmer* was it in the afternoon *than* in the morning?

S7. *Two unequal amounts are given.* They are to be compared by finding their *difference.*

Expressions characteristic of this pattern are *difference in* and *difference between.*

(1) Tom paid 25¢ for a ticket to the school play. Bobby paid 15¢ for his ticket. What is the *difference between* the prices of these tickets?

(2) Betty is 9 years old. Her mother is 31 years old. What is the *difference between* their ages?

S8. *Two unequal amounts are given.* They are to be compared by finding their difference but the word *difference* is not used in the problem.

There are no technical expressions that are characteristic of the problems of this pattern. In these problems the answer, which represents a difference, is subject to a fair range of interpretation. This pattern is often met in problems where the difference represents a gain or a loss, or an increase or a decrease. In problems of this pattern one must look for the fundamental thought in each problem as a guide to its solution.

(1) Alice weighed 81 pounds last year. Now she weighs 87 pounds. How many pounds has she gained?
(2) The first telephone was invented in 1876. How many years ago was that?
(3) The temperature early this morning was 58°. At noon it was 67°. How much did the temperature rise?
(4) The speedometer on Mr. Hart's car read 987.6 mi. on Sunday and 1177.5 mi. at the end of the week. How many miles did he drive that week?

S9. *Two unequal amounts have been compared. One* of the amounts and the *difference between* them are given. Find the *other* amount. This is the inverse of S5, S6, S7, and S8. No inverse of S6 or S7 was found in the texts. Pattern S9 is concerned with the comparison of two numbers but in this case the comparison has already been made; that is, the *difference between* the two numbers and *one* of the numbers are known and one works backward, as it were, to find the *other* number.

(1) Alice has 10 jacks. Grace says, "Alice has 2 more than I have." How many does Grace have?
(2) Abraham Lincoln died in 1865. He was 56 years old. In what year was he born?

The fourth group contains only one pattern, which differs fundamentally from the other three groups of patterns, as will be seen from the following discussion and illustrations.

IV. THE PATTERN THAT INVOLVES THE "SEPARATION-INTO-PARTS" IDEA

S10. Given the sum and one part, find the other part.

In this pattern, there is no decrease of a given amount, no finding of the amount left, and no comparison. The concern is merely to find one part when the sum and the other part are given.

(1) There are 27 children in Mary's grade when all the children are present. Today only 24 children are present. How many are absent?

(2) Ned has 245 hens. He has 186 brown hens and the rest are white. How many white hens has he?

(3) The school nurse weighed 845 children. Of this number 269 were underweight. How many of the children were not underweight?

(4) When Betty and Joe were weighed together, they weighed 153 pounds. If Joe alone weighed 81 pounds, how many pounds did Betty weigh?

It will be seen that the question in each problem of Pattern S10 is peculiar to that problem. There is no one question characteristic of the pattern as a whole. The characteristic common to all problems of this pattern is that the *sum* of two numbers and *one* of them are known, and the question is concerned with finding the *other* number. Looked at in the large, this is the general definition of subtraction and really includes all the other subtraction patterns. In this study Pattern S10 is to be applied in its narrower sense and is not to be considered a substitute for all the other subtraction patterns.

The ten one-step subtraction patterns listed above were determined by examining all the one-step subtraction problems and the subtraction step in the multiple-step problems of the four series of texts, covering Grades 3 through 6, which were analyzed in this study. This examination revealed 2,998 one-step subtraction elements, of which 2,178 came from one-step problems and 820 from multiple-step problems. It is both interesting and significant that all of these 2,998 one-step subtraction elements could be grouped into ten patterns. Ten patterns is a very small number, considering the wide variety of social situations and language expressions used in these problems. Yet, in spite of the variation in social situations and language expressions, certain likenesses are discernible which make it possible to break down these problems into ten patterns.

As each one-step problem belonging to a specific pattern contributes to the formulation of a generalization about that pattern, so each pattern contributes to the formulation of certain larger generalizations regarding the process of subtraction. Finally, subtraction comes to mean the process which is used (1) when a given amount has been decreased and the amount left is required, (2) when a given amount is on hand and it is required to find how much more is needed to equal a desired total amount, (3) when two numbers are to be compared by finding their difference, and (4) when the whole and one part are given and it is required to find the other part. Only after wide experience with many problems of each pattern and recognition of the common elements in each pattern can these broad generalizations regarding subtraction be formulated. Understanding of these generalizations is an important goal in problem solving.

Distribution of the Ten Subtraction Patterns in Four Series of Arithmetics

Before beginning the study of the distribution of the subtraction patterns in the four series of texts, it will be necessary to explain the terms *hidden step* and *final step* which are used in connection with the analysis of problems of two or more steps. The following two-step problem will illustrate these terms: "At the school store John bought a pencil for 5 cents and a pad for 8 cents. He gave the clerk 25 cents. How much change did he receive?" This is a two-step problem since it requires two operations to get the answer; the first operation is the addition of 5 and 8 and the second operation is the subtraction of 13 from 25. The first step in this solution is called a *hidden step* since there is nothing in the problem that helps the pupil to decide that 5 and 8 are to be added. He does get some help, however, in the second step, which is called the final step; this help came from the question *how much change did he receive?* It is evident that every two-step problem contains one hidden step and one final step, whereas a three-step problem contains two hidden steps and one final step. To solve a one-step problem requires merely the application of some one pattern. To solve a three-step problem requires the application of three one-step patterns, two of these patterns being applied in the hidden steps and one in the final step. In analyzing all the subtraction problems in the four series of books used in this study, each problem was broken down into its one-step patterns and a record was made as to whether the pattern was applied in a one-step problem or a multiple-step problem and, if applied in a multiple-step problem, a record was made as to whether the application was in a hidden step or in the final step.

It will be of interest to know to what extent each of the subtraction patterns is repeated in the problem material provided in each of the four series of books. While it is not a part of this study to state how many repetitions of each pattern are necessary, it is possible to determine whether inconsistent practices exist with respect to these repetitions. It is assumed, in this study, that if a pupil does not understand the underlying thought in certain of the one-step patterns, he cannot successfully solve problems that involve those patterns. A certain amount of repetition of each pattern is necessary in order that the pupil may become familiar with the basic thought underlying it.

In the discussions in this chapter the four series of books analyzed in this study will be referred to as Series X, Series Y, Series Z, and Series W. For brevity the terms Text X, Author X, or merely X may at times be used to designate Series X; similarly, Text Y, Author Y, or Y will refer to Series Y, and so on.

Subtraction Pattern S1

All the one-step patterns discussed in this study are illustrated in graphic charts. The chart for Subtraction Pattern S1 appears on page 25. A few suggestions will aid in the interpretation of these charts. In the chart for Pattern S1 and, in particular, the graph for Grade 4 on that chart, it is seen that the first horizontal bar, which represents Text X, is divided into two parts; the "open" part of the bar (the part at the left) shows the number of repetitions of Pattern S1 that were found in one-step problems; there were 9 such repetitions. The "solid" part of the bar (the part at the right) shows the number of repetitions of Pattern S1 that occurred in multiple-step problems; there were 30 such repetitions. The number at the left of the bar represents the *total number of repetitions* of Pattern S1 in both one-step and multiple-step problems. The "solid" part of each bar includes those repetitions of Pattern S1 that occur both in the hidden steps and in the final steps of the multiple-step problems.[1] For the purposes of this chapter it was not found necessary to subdivide the "solid" part of each bar into two smaller parts to distinguish the hidden steps from the final steps.[2] In the discussions there will be need to refer only to two items represented by each bar, namely, the number of repetitions of each pattern found in one-step problems (represented by the "open" part of the bar) and the total number of repetitions of that pattern as found in one-step and multiple-step problems combined (represented by the entire length of the bar). For brevity the terms "one-step repetitions" and "total repetitions" will be used to indicate these two types of repetitions.

[1] It should be pointed out that the "solid" part of each bar does not represent the number of multiple-step problems in a given text but the number of repetitions of Pattern S1 that are found in multiple-step problems. A single problem of the three-step variety represents three patterns, of which one might be a multiplication pattern, one an addition pattern, and one a subtraction pattern. In tabulating the results of the analysis of this three-step problem, each of the three patterns concerned would be credited with one occurrence.

[2] If one is interested to know the frequency of the hidden steps and final steps in multiple-step problems, this information will be found in Tables XIVa–XIVd in the Appendix, pp. 159 ff.

A more detailed study of the graphs of Pattern S1 will reveal some interesting facts. Consider first the graph for Grade 3. Here we find that Text X has a total of 63 repetitions of S1, all of which are in one-step problems. Text Y has a total of 52 repetitions, of which 51 are in one-step problems; only 1 repetition occurs in a multiple-step problem. The bars of Texts Z and W explain themselves. One striking feature in the graph for Grade 3 is that two texts, X and W, exercise Pattern S1 exclusively in one-step problems; similarly, the other two texts, Y and Z, have all but 1 and 2 repetitions respectively in one-step problems. It is evident that repetitions of Pattern S1 in multiple-step problems are a negligible quantity so far as Grade 3 is concerned.

If we study critically the graph for Grade 4 we find that the four texts are in fairly close agreement with respect to the total number of repetitions of Pattern S1 in this grade, these totals varying from 36 to 43. A new feature in Grade 4 is the liberal appearance of Pattern S1 in multiple-step problems, Text X having the greatest number of such repetitions. Another new feature in this grade is the wide variation among authors in the offering of one-step problems. Text X has only 9 one-step problems, while each of the other texts has three or more times as many. The relative emphasis that an author gives to one-step problems as compared with multiple-step problems will receive attention from time to time as this discussion develops since it seems reasonable to expect that a pattern should have considerable repetition in one-step problems before moving to multiple-step situations. It has been observed, however, that one-step problems have been generously supplied in Grade 3.

The graph for Grade 5 presents a situation similar to that in Grade 4 but with somewhat less agreement in Grade 5 on the "total repetitions" of Pattern S1 and far less agreement on the repetitions in one-step situations. It is also evident that Pattern S1 is exercised more frequently in Grades 3 and 5 than it is in Grades 4 and 6.

The graph for Grade 6 explains itself. The graph for *all*

four grades combined shows very close agreement among Texts X, Y, and Z with reference to their total offerings of Pattern S1. Text W has the largest offering of the four texts.

It will be found as this study progresses that the treatment of Pattern S1 in each of the four grades is more uniform with respect to the four texts than is the case for most of the other patterns of the subtraction group. This uniformity, of course, refers to the total number of repetitions of each pattern. There is considerable variation, however, with respect to the treatment of one-step problems in each grade. In order to have a convenient means of comparing the variation of the treatment of the four authors in each grade use is made of the coefficient of variation, represented by V, which is defined as 100 sigma divided by the mean. This coefficient of variation has been computed for each grade for the "total repetitions" and the "one-step repetitions" of each pattern. Thus, in Grade 4 of the chart for Pattern S1, it is seen that $V=6.5$ for "total repetitions," while $V=38.7$ for "one-step repetitions." It is evident that the variation is far greater for "one-step repetitions" than for "total repetitions."

Subtraction Pattern S2

The graphic chart for Pattern S2, which is concerned with the amount of change that is received when a purchase is made, is shown on page 26. In this chart it is seen that there is considerable variation in each grade in the offerings of the four authors. In a way this variation with respect to Pattern S2 is not so serious as it might seem at first since a child familiar with Pattern S1 might be expected to transfer some of his ability to problems of Pattern S2. Further, Pattern S2 arises in only one type of situation, namely, that in which a purchase is made, whereas Pattern S1 can arise in any situation where a given amount has been decreased. The reason for keeping S2 as a separate pattern rather than merging it with S1 is that retail buying is a social activity that concerns all of us.

A study of Pattern S2 will now be made. In Grade 3 there is wide variation among the authors in the "total repetitions"

SUBTRACTION PATTERNS

SUBTRACTION PATTERN S1

Finding How Much Is Left

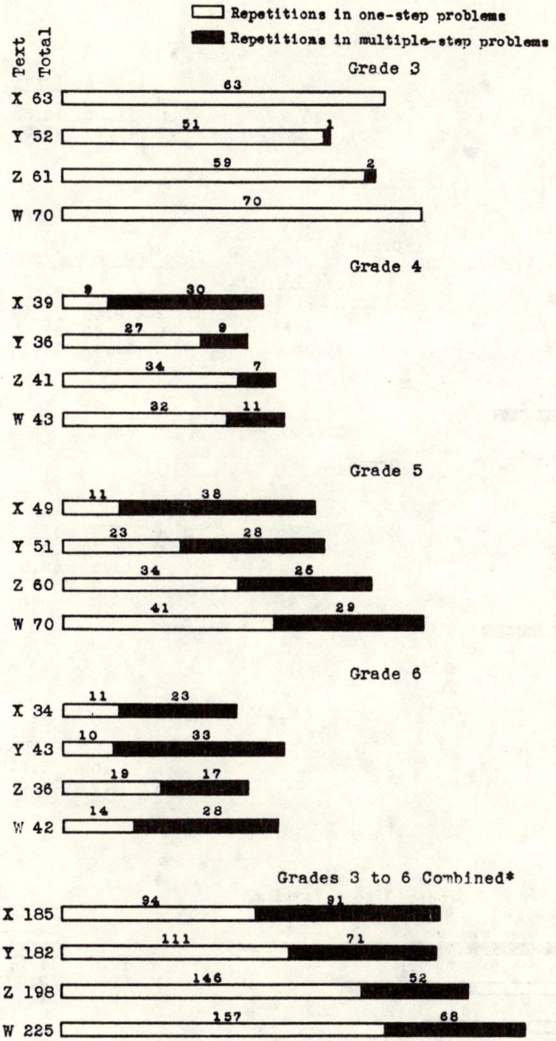

Grade 3

	1-Step	Total
σ	6.87	6.42
M	60.75	61.50
V	11.3	10.4

Grade 4

	1-Step	Total
σ	9.86	2.59
M	25.50	39.75
V	38.7	6.5

Grade 5

	1-Step	Total
σ	11.37	8.32
M	27.25	57.50
V	41.7	14.5

Grade 6

	1-Step	Total
σ	3.50	3.83
M	13.50	38.75
V	25.9	9.9

Grades 3 to 6 Combined

	1-Step	Total
σ	25.42	16.98
M	127.00	197.50
V	20.0	8.6

*The graph for Grades 3 to 6 Combined has been drawn to a reduced scale.

SUBTRACTION PATTERN S2

Finding How Much Change Is Received After a Purchase

Grade 3

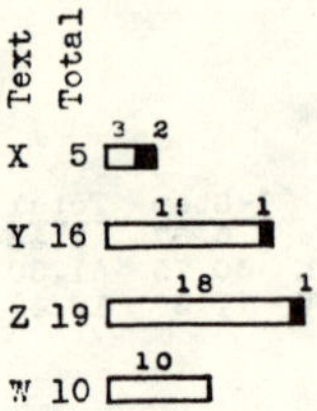

	1-Step	Total
σ	5.68	5.31
M	11.50	12.50
V	49.4	42.5

Grade 4

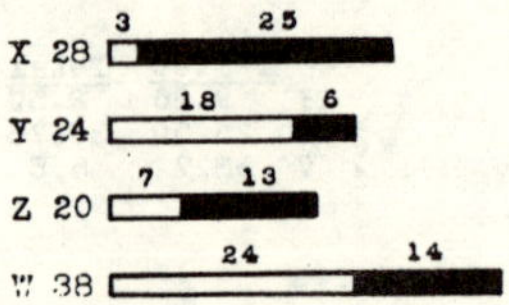

	1-Step	Total
σ	8.40	6.69
M	13.00	27.50
V	64.6	24.3

Grade 5

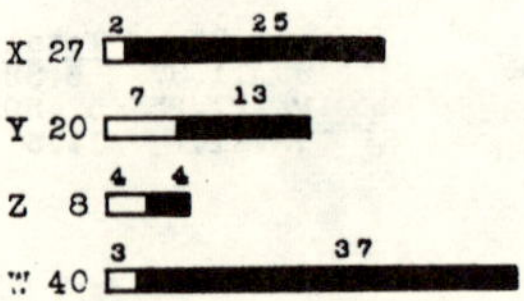

	1-Step	Total
σ	1.67	11.05
M	4.00	23.75
V	46.8	46.5

Grade 6

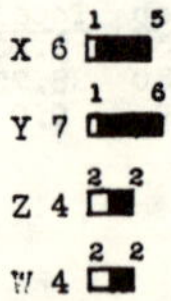

	1-Step	Total
σ	0.50	1.30
M	1.50	5.25
V	33.3	24.8

Grades 3 to 6 Combined

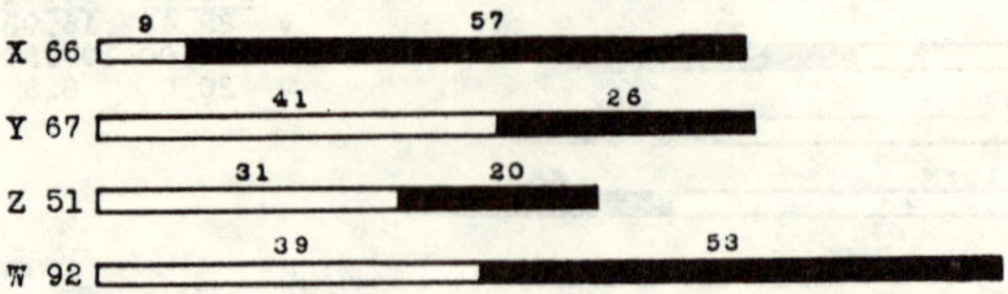

	1-Step	Total
σ	12.69	14.71
M	30.00	69.00
V	42.3	21.3

SUBTRACTION PATTERN S3

Given the Minuend and the Remainder, to Find the Subtrahend

Grade 3

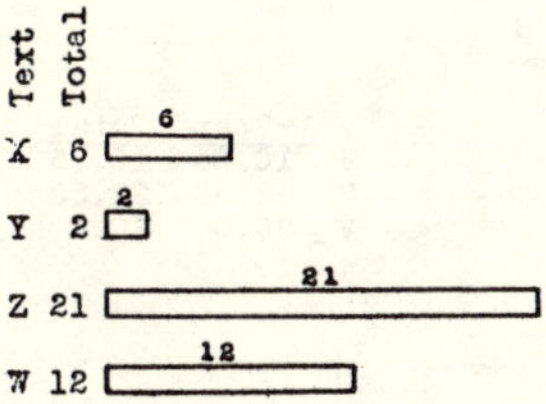

	1-Step	Total
σ	7.15	7.15
M	10.25	10.25
V	69.8	69.8

Grade 4

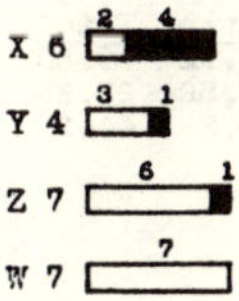

	1-Step	Total
σ	2.06	1.22
M	4.50	6.00
V	45.8	20.3

Grade 5

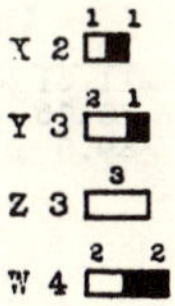

	1-Step	Total
σ	0.71	0.71
M	2.00	3.00
V	35.5	23.7

Grade 6

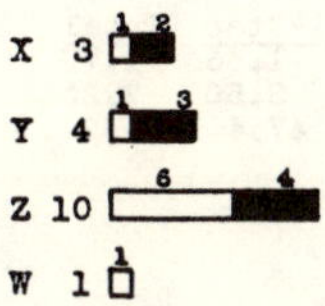

	1-Step	Total
σ	2.17	3.35
M	2.25	4.50
V	96.4	74.4

Grades 3 to 6 Combined

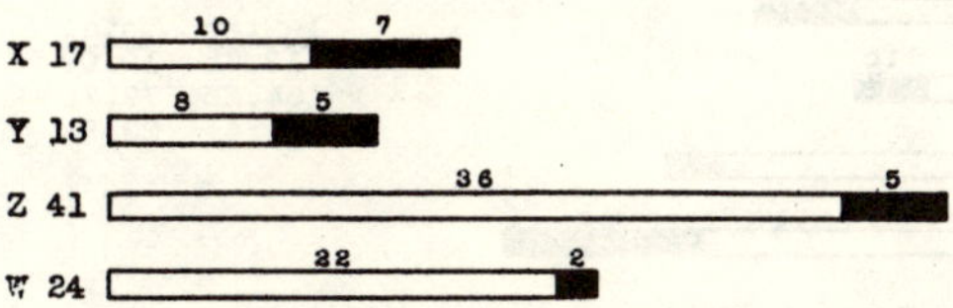

	1-Step	Total
σ	11.18	10.71
M	19.00	23.75
V	58.8	45.1

SUBTRACTION PATTERN S4

Finding How Much More Is Needed to Reach a Desired Amount

Grade 3

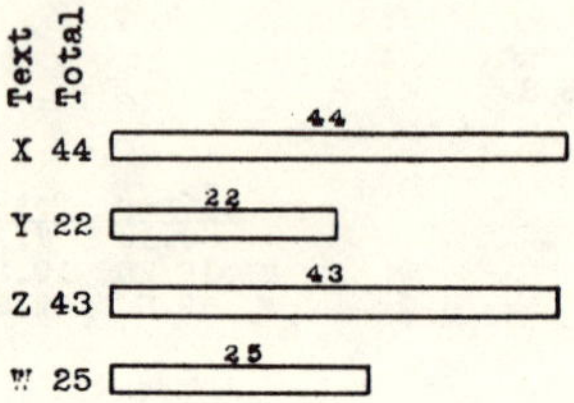

	1-Step	Total
σ	10.06	10.06
M	33.50	33.50
V	30.0	30.0

Grade 4

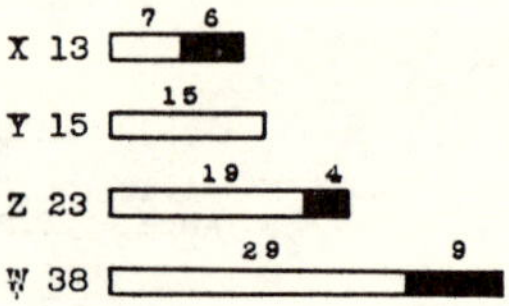

	1-Step	Total
σ	7.92	9.83
M	17.50	22.25
V	45.3	44.2

Grade 5

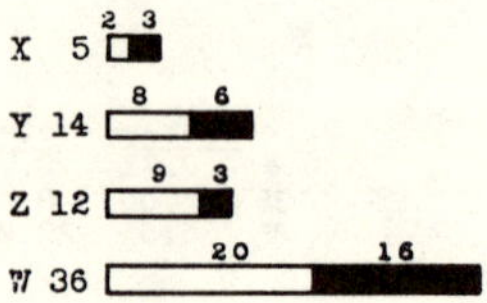

	1-Step	Total
σ	6.50	11.61
M	9.75	16.75
V	66.7	69.3

Grade 6

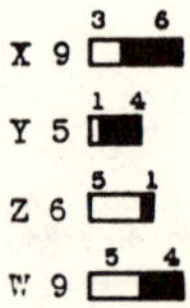

	1-Step	Total
σ	1.66	1.79
M	3.50	7.25
V	47.4	24.7

Grades 3 to 6 Combined*

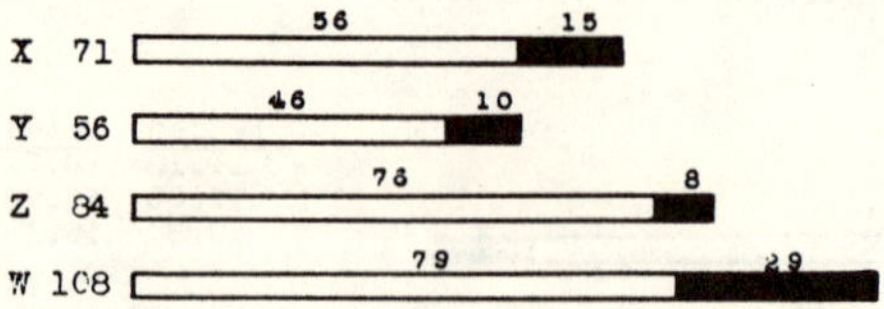

	1-Step	Total
σ	13.75	19.08
M	64.25	79.75
V	21.4	23.9

*The graph for Grades 3 to 6 Combined has been drawn to a reduced scale.

SUBTRACTION PATTERN S5

Comparing Two Numbers, Using the Phrases
How Much More, or *How Much Less*

Grade 3

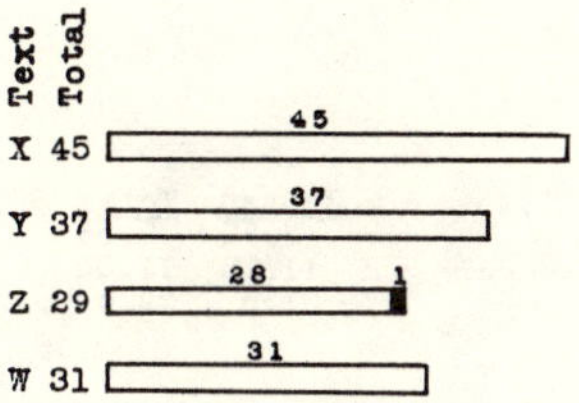

	1-Step	Total
σ	6.50	6.22
M	35.25	35.50
V	18.4	17.5

Grade 4

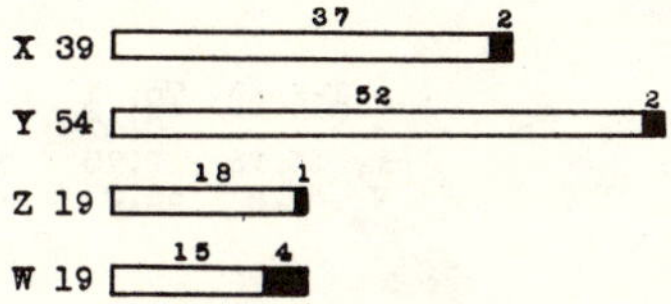

	1-Step	Total
σ	15.00	14.74
M	30.50	32.75
V	49.2	45.0

Grade 5

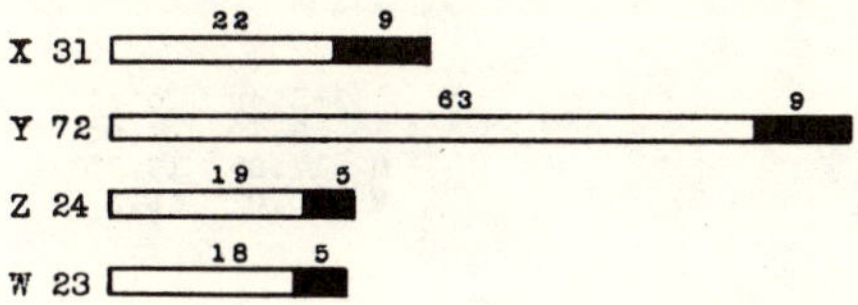

	1-Step	Total
σ	18.82	20.16
M	30.50	37.50
V	61.7	53.8

Grade 6

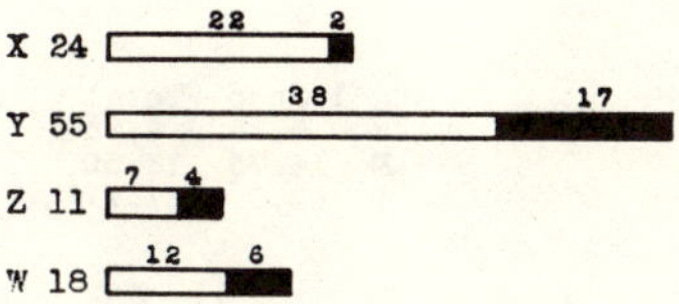

	1-Step	Total
σ	11.84	16.81
M	19.75	27.00
V	59.9	62.3

Grades 3 to 6 Combined*

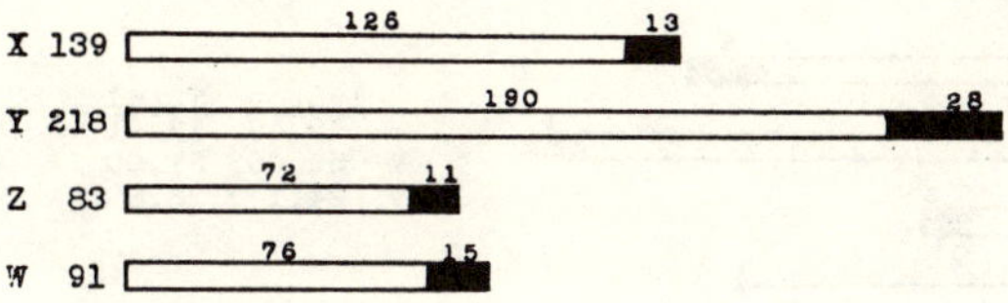

	1-Step	Total
σ	47.73	53.68
M	116.00	132.75
V	41.1	40.4

*The graph for Grades 3 to 6 Combined has been drawn to a reduced scale.

SUBTRACTION PATTERN S6

Comparing Two Numbers, Using the Phrases *How Much Larger, How Much Older,* Etc.

Grade 3

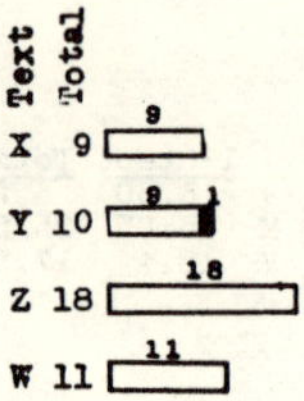

	1-Step	Total
σ	3.70	3.54
M	11.75	12.00
V	31.5	29.5

Grade 4

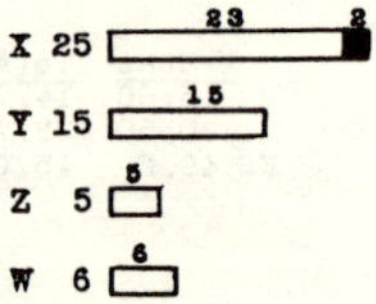

	1-Step	Total
σ	7.33	8.07
M	12.25	12.75
V	59.8	63.3

Grade 5

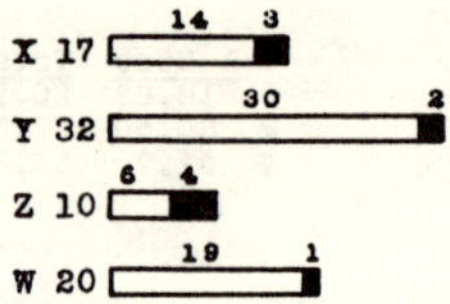

	1-Step	Total
σ	8.70	7.95
M	17.25	19.75
V	50.0	40.3

Grade 6

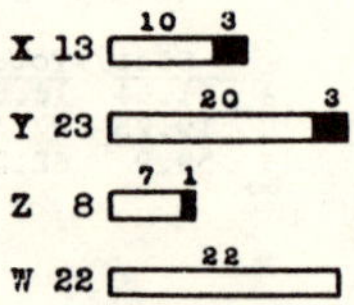

	1-Step	Total
σ	6.38	6.26
M	14.75	16.50
V	43.3	37.9

Grades 3 to 6 Combined

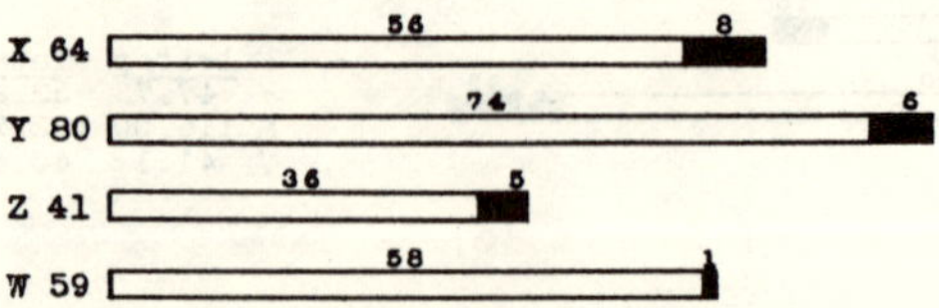

	1-Step	Total
σ	13.49	13.91
M	56.00	61.00
V	24.1	22.8

SUBTRACTION PATTERN S7

Comparing Two Numbers, Using the Word *Difference*

Grade 3

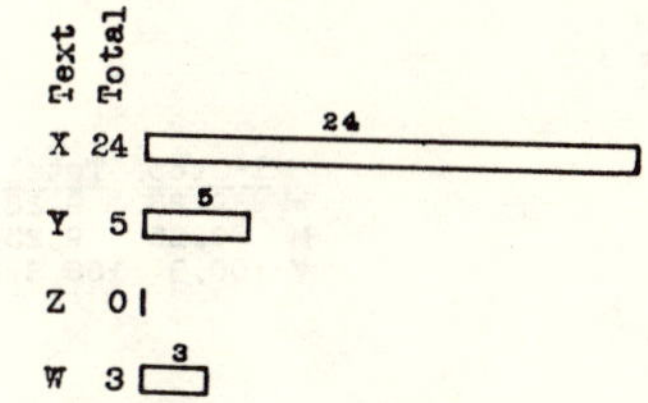

	1-Step	Total
σ	9.41	9.41
M	8.00	8.00
V	117.6	117.6

Grade 4

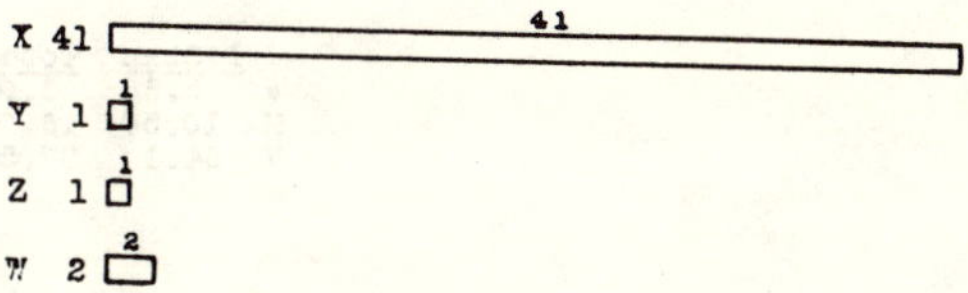

	1-Step	Total
σ	17.18	17.18
M	11.25	11.25
V	152.7	152.7

Grade 5

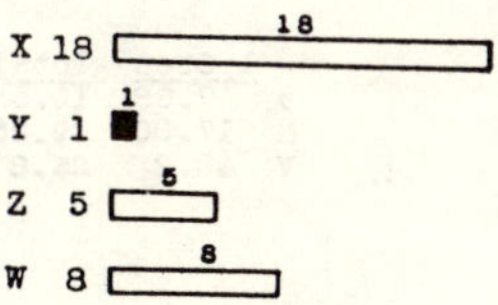

	1-Step	Total
σ	6.57	6.28
M	7.75	8.00
V	84.8	78.5

Grade 6

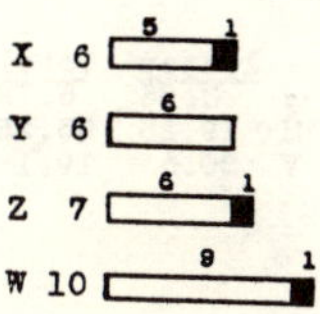

	1-Step	Total
σ	1.50	1.64
M	6.50	7.25
V	23.1	22.6

Grades 3 to 6 Combined*

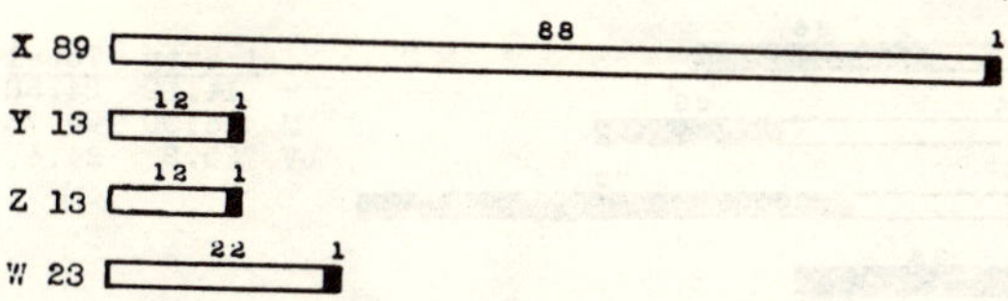

	1-Step	Total
σ	31.73	31.73
M	33.50	34.50
V	94.7	92.0

*The graph for Grades 3 to 6 Combined has been drawn to a reduced scale.

SUBTRACTION PATTERN S8

Comparing Two Numbers, Using No Characteristic Phraseology

Grade 3

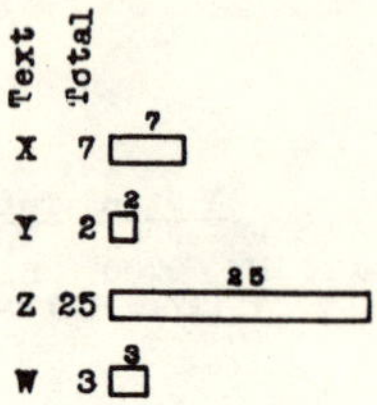

	1-Step	Total
σ	9.28	9.28
M	9.25	9.25
V	100.3	100.3

Grade 4

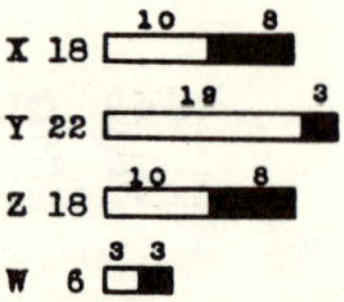

	1-Step	Total
σ	5.68	6.00
M	10.50	16.00
V	54.1	37.5

Grade 5

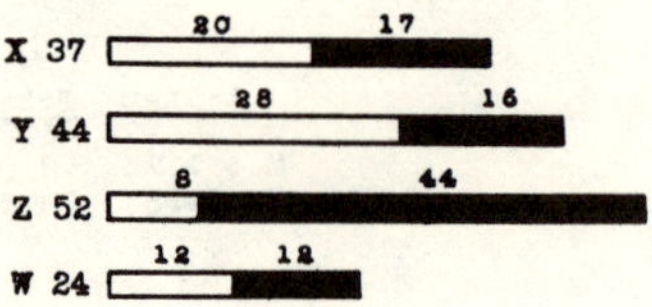

	1-Step	Total
σ	7.68	10.52
M	17.00	39.25
V	45.2	26.8

Grade 6

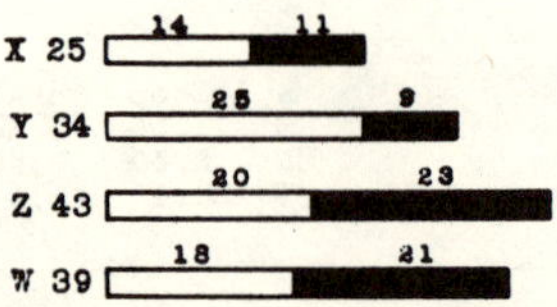

	1-Step	Total
σ	3.96	6.72
M	19.25	35.25
V	20.6	19.1

Grades 3 to 6 Combined*

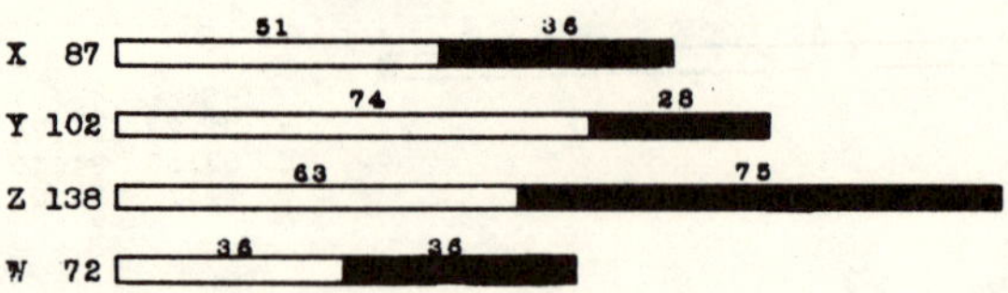

	1-Step	Total
σ	14.12	24.50
M	56.00	99.75
V	25.2	24.6

*The graph for Grades 3 to 6 Combined has been drawn to a reduced scale.

SUBTRACTION PATTERN S9

Inverse of Patterns S5, S6, S7, and S8

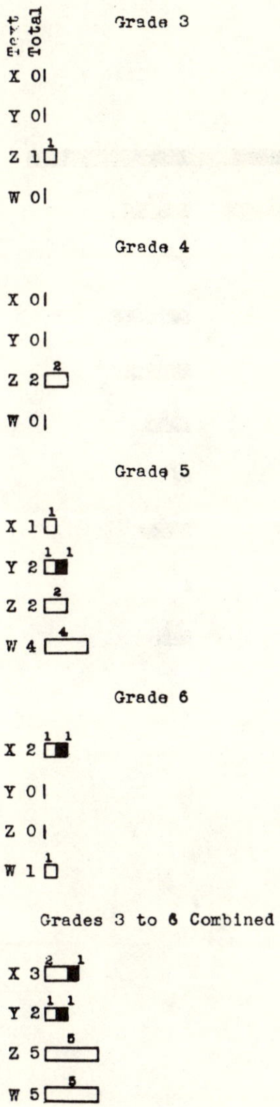

SUBTRACTION PATTERN S10

Given the Total and One Part, to Find the Other Part

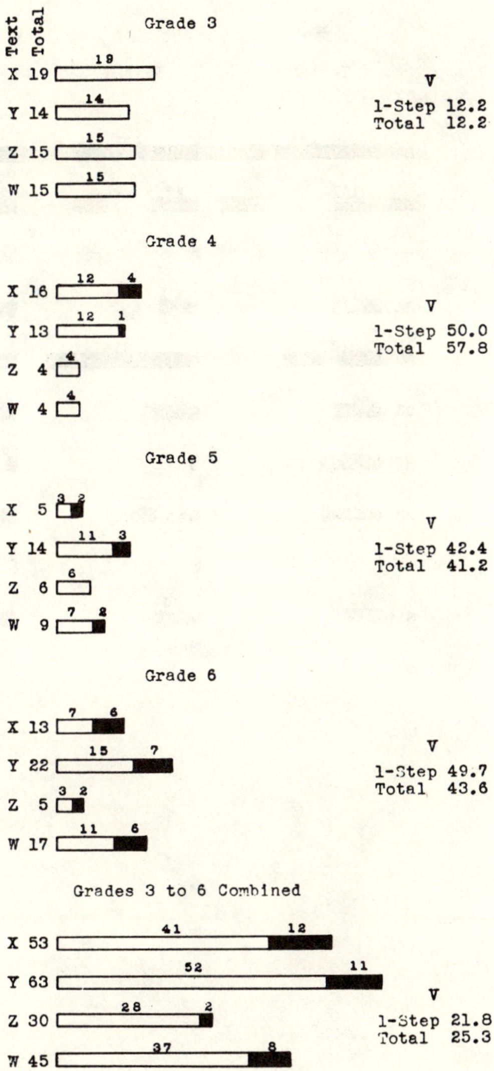

CHART 1

Total Repetitions of Each of the Subtraction Patterns in All Four Grades Combined, Arranged by Texts

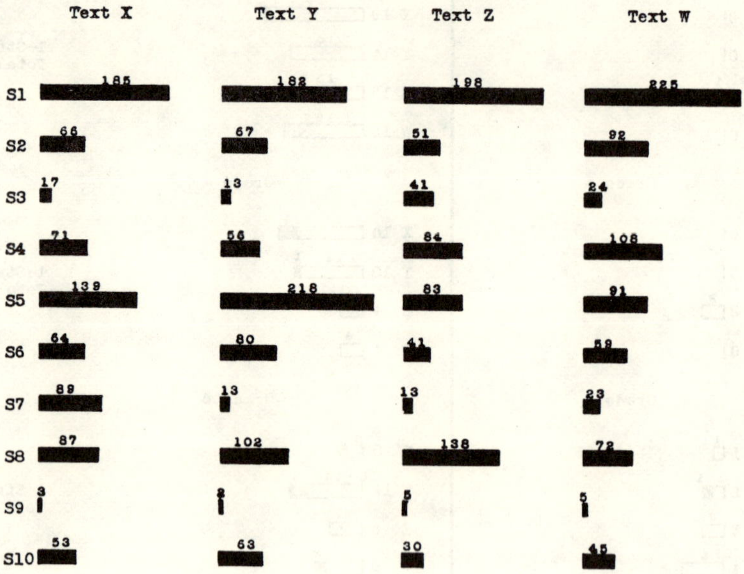

of this pattern. Author X gives 5 repetitions, whereas Author Z gives almost four times as many.

The graphs show that in Grades 4 and 5 Pattern S2 appears frequently in multiple-step problems; in fact, this pattern is usually in its most natural situation in a multiple-step problem. In Grade 5 there is wide variation in the "total repetitions" of Pattern S2, as is seen by the fact that Text W has five times as many repetitions as Text Z and twice as many repetitions as Text Y. It seems probable that Author Z would have increased his total offering in Grade 5 if he had been aware of the facts disclosed by this graph. The wide variation of "total repetitions" in Grade 5 is indicated by a value of 46.5 for V.

In Grade 6 there is a marked falling off in Pattern S2 in all four texts. In the graph for *all four grades combined* the discrepancies in the separate grades are evened up somewhat with reference to "total repetitions," though there is still wide variation, represented by $V = 42.3$, with respect to the "one-step repetitions." When the treatment of Pattern S2 is considered as a whole, it would not be unreasonable to expect a more liberal offering of this pattern with a more even distribution of it over the four grades, particularly the last two, since this pattern is related to the important life activity of making purchases.

Subtraction Pattern S3

The graphic chart of Pattern S3 is given on page 27. From this chart it is evident that this pattern receives very little attention in any of the four grades. Examining the total number of appearances in all four grades combined, as shown in the graph, one finds that this pattern receives only 13 repetitions in Text Y and that the maximum treatment is 41 repetitions in Text Z. In each case this is a very small offering distributed over a period of four years. Pattern S3 belongs to the group representing the remainder idea and is the inverse of Patterns S1 and S2. This is the pattern where the minuend and the remainder are given and the subtrahend has to be found. It is illustrated by the following problem: "John had 12 rabbits but some of them ran away. He has 7 rabbits left. How many rabbits ran away?" Inverse cases are generally considered more difficult than direct cases. Inverse cases that come under Pattern S3, however, are fairly easy to understand since they can easily arise in rather concrete life situations. The criticism is sometimes raised that inverse cases, in general, do not represent real problems, but such criticism is not justified in the case of Pattern S3. There are many occasions where the *amount left* is known but the *amount taken away or spent* has been forgotten and it is necessary to work backward to find the desired fact. Another illustration of this pattern is the following: "This morning I had $5.00 and this evening I have $2.15 left in my purse. How much did I spend?" Pattern S3 is also applied when one balances a cash book by counting the cash on hand and subtracting that amount from the total receipts to see if the balance agrees with the total expenditures. In view of the broad usefulness of Pattern S3 it would seem reasonable to expect that this pattern should receive considerably more emphasis than it has in each of the four grades.

Subtraction Pattern S4

The chart for Pattern S4 is given on page 28. Pattern S4 represents the second important idea of subtraction, namely, that of building up a given amount until it equals a larger desired total.

Referring to the graphic chart for Pattern S4, it is found in each grade that there is wide variation in the offering of the four authors. In Grade 4, for example, Text W provides 38 "total repetitions" of this pattern, which is about *three times* the number offered by Text X. In Grade 5 the extremes in the offering are still greater, ranging from 5 in Text X to over *seven times* as many in Text W. These wide variations in the individual grades are not evened up when one considers the graph of the combined offering for the four grades. In this combined graph Text Y has 56 repetitions, whereas Text W has 108 repetitions, or about twice as many.

Pattern S4 is an important one. Out of the ten patterns in

subtraction this pattern ranks second in Text W, third in Text Z, and fifth in Text X, these facts being shown in the table on page 44. If one takes into consideration the practical importance of the building-up idea in subtraction, it seems probable that the authors of those texts that have few repetitions of Pattern S4 in certain grades would increase their offering of this pattern if they were aware of the facts stated above.

Subtraction Patterns S5 and S6

The third important group of subtraction patterns is concerned with the comparison of two numbers. This comparison group is made up of five patterns, namely, S5, S6, S7, S8, and S9. Let us first consider Patterns S5 and S6, which are very closely related. In Pattern S5 the comparison is indicated by asking the question in the form *how much more* or *how much less,* whereas in S6 the comparison is indicated by such expressions as *how much longer, how much shorter, how much larger, how much heavier, how much older, how much younger,* etc. It is evident that Pattern S6 may be expressed in a wider range of phraseology than S5; in fact, Pattern S5 is confined to the two phrases *how much more* and *how much less.* Both these patterns, S5 and S6, have wide practical application. The graphic chart of Pattern S5 is given on page 29. In Grade 3 of this chart there is fair uniformity among the four authors in the "total repetitions" of this pattern. In Grades 4, 5, and 6, however, there is wide variation among the authors in the "total repetitions," the coefficients of variation being respectively 45.0, 53.8, and 62.3. It is evident in Grades 4, 5, and 6 that Author Y is very partial to Pattern S5. In Grade 5 Author Y has three times as many "total repetitions" as Author Z or Author W, while in Grade 6 Author Y has five times the offering of Author Z. In the combined graph for all four grades, Author Y has 218 "total repetitions," which is considerably more than twice the offerings of Authors Z and W.

The graphic chart for Pattern S6 is given on page 30. In the combined graph at the bottom of this chart, covering all four grades, one finds that each author gives Pattern S6 considerably less attention than was given S5; this is a rather striking difference since the verbal expressions peculiar to Pattern S6 are much more varied than those of S5. One might naturally have expected the emphasis to be in favor of Pattern S6 rather than S5. It is conceded, however, that Pattern S5 has wide social application. In view of the practical utility of Patterns S5 and S6 it seems probable that some of the four authors, if their attention were called to this situation, would prefer to redistribute their problems so as to increase inadequate offerings and, in the case of Author Y, probably decrease some of the excessive offering of Pattern S5. Another interesting fact disclosed by the graphs of Patterns S5 and S6 is the liberal number of repetitions of these patterns in one-step problems as compared with their appearance in multiple-step problems.

Subtraction Pattern S7

Pattern S7 also represents the comparison idea of subtraction. The technical language of Pattern S7 is limited to those expressions involving the word "difference," such as *find the difference* or *find the difference between.* If we examine the graphic chart for S7, given on page 31, we will find that only one author, X, gives much attention to this pattern. The variation within each grade of the offerings of the four authors is very great. There are several extreme situations involving this pattern. In Grade 3 Text Z omits Pattern S7 entirely and in Grade 4 gives it only once. Likewise, Text Y gives this pattern only once in Grade 4 and once in Grade 5. In the combined graph for all four grades, the offering of Text X is almost seven times as great as that of Texts Y or Z and about four times as great as that of Text W. It is hard to believe that this pattern would be neglected by certain authors if those authors were aware of the inadequacy of their offerings. An observation of further interest is that among all the authors such repetitions of this pattern as are given are mostly in one-step problems; multiple-step problems involving this pattern are negligible.

The meager treatment of Pattern S7 given by three of the four authors may be an indication that it would be better to omit Pattern S7 in Grade 3, thus leaving its first appearance to Grade 4. The word *difference* and the concept *difference* are both difficult for young pupils; further, there are so many other subtraction patterns in Grade 3 that it might be well to defer the more difficult ideas. It is the recommendation of the writer that Pattern S7 be delayed until Grade 4. To delay this pattern until Grade 4 does not, of course, excuse its neglect in Grades 4, 5, and 6 on the part of Authors Y, Z, and W.

Subtraction Pattern S8

Pattern S8 belongs to the comparison group. This pattern requires a difference to be found, but the word *difference* is not used in the problems. This pattern has a wide range of technical phraseology associated with it. It is often involved in problems where one is required to find a gain or a loss, or an increase or a decrease, or a distance traveled if one has the speedometer readings before and after a trip. In other words, S8 is a very useful pattern and sufficiently difficult in certain of its technical applications to justify ample treatment on the part of authors.

The graphic chart for Pattern S8 is given on page 32. In Grade 3 it is seen that there is wide variation among the authors in their offerings of this pattern. One author gives it 25 repetitions while the other three authors give it only 7, 2, and 3 repetitions respectively. One, no doubt, would agree that these four authors are not each correct. If we study the offerings in Grades 4, 5, and 6, we find that they are more generous, particularly in Grades 5 and 6, with liberal application in multiple-step problems. This greater emphasis on Pattern S8, particularly in the intermediate grades, is no doubt in harmony with the social utility of this pattern. Further, the technicalities of this pattern are sufficiently difficult to justify generous repetition, particularly in Grades 4, 5, and 6. It is noticeable from the graphs of Grades 4, 5, and 6, and also from the coefficients of variation, that there is greater variation in each grade in the offering of this pattern in one-step form than in multiple-step form.

It was pointed out above that Pattern S8 has very erratic representation among the four authors in Grade 3. No doubt this wide variation reflects the difficulty of this pattern and the hesitancy on the part of the authors to introduce it too soon. It might be worth while seriously to consider the advisability of omitting Pattern S8 in Grade 3, leaving it until Grade 4 for its initial appearance. A similar recommendation was made above for Pattern S7; the reasons for suggesting the postponement of Pattern S8 until Grade 4 are the same as those given with respect to the delay of Pattern S7.

Subtraction Pattern S9

Pattern S9 is the last pattern of the comparison group. This pattern is the inverse of comparison Patterns S5, S6, S7, and S8. Illustrations of this pattern are given below:

(1) Tom weighs 112 lb. which is 8 lb. more than Joe weighs. How many pounds does Joe weigh?
(2) Mary painted 7 more Christmas cards today than yesterday. Mary painted 29 cards today. How many did she paint yesterday?
(3) The speedometer on Fred's bicycle read 179 mi. when he reached home this evening. During the day he rode the bicycle 7 mi. What did the speedometer read this morning?

A glance at the graphic chart of Pattern S9, given on page 33, shows that this pattern is almost non-existent so far as the four texts under consideration are concerned. In all four grades combined this pattern appears only 3 times in Text X, 2 times in Text Y, and 5 times each in Texts Z and W. In other words, Pattern S9 occurs *only 15 times in four years* in all four texts combined, which is roughly an average of about *one appearance per year per text.* What a meager showing this is! One is at a loss to know whether the authors were unaware of this pattern or indifferent to it. One infers the former. A hasty judgment might lead some to say that a problem of this pattern is unreal, that the answer has to be known before such a problem can be stated. Whether such problems are real or not depends largely upon the circumstances. In the first illustrative problem above, it is true that someone had to know Joe's weight before it could be determined that Tom weighed 8 lbs. more than Joe. The person who originally knew Joe's weight may have been the teacher at school and it was at school that Tom discovered that he weighs 8 lbs. more than Joe. When Tom goes home, however, he remembers only his own weight, 112 lbs., and the fact that he weighs 8 lbs. more than Joe. Hence the problem is a real one when Tom gets home if some member of his family wants to know how much Joe weighs. There are many problems in life where the original facts are known to someone, but some of the original data may be missing when these facts get to a second person. This confronts the second person with a problem of seeing if he can find the missing data *indirectly*. It is in this way that all indirect or inverse cases arise.

The writer grants that inverse cases represented by Pattern S9 arise much less frequently than the direct cases represented by Patterns S5, S6, S7, and S8. This fact, however, does not seem to justify the neglect of Pattern S9 on the part of all the authors which is so evident in the graphic chart on page 33. It might be well to delay the introduction of Pattern S9 until Grade 4, particularly if Patterns S7 and S8 are postponed until Grade 4, as has already been recommended.

Subtraction Pattern S10

The fourth important idea of subtraction, that of finding *one part* when the *whole* and the *other part* are given, is represented by Pattern S10. This pattern is illustrated by the following problem: "There are 27 pupils in Miss White's room at school. 14 of them are boys; how many girls are there?" It was pointed out in the early part of this chapter, where Pattern S10 was first described, that subtraction might be defined as an operation in which one is given the *sum* of two numbers and *one* of the numbers and is required to find the *other* number. This definition is sufficiently broad to include any of the patterns already described, including those patterns representing the remainder idea and the comparison idea. Pattern S10, however, is not considered in this study as representing this general definition of subtraction. Pattern S10 is limited to those cases, like the problem given above, where the *total* and *one of the parts* is given and the *other part* is to be found. In applying this pattern to the problems that represent it, there is no consciousness on the part of the pupil of comparing numbers, or of finding how much is left; the pupil is merely concerned to know the size of the unknown part.

The findings concerning Pattern S10 are shown in graphic form on page 33. In Grade 3 it is seen that the treatment of this pattern is rather uniform among all four authors, though not overgenerous in amount, but in Grades 4, 5, and 6 there is considerable variation in the offerings, with a marked decrease in the offerings of certain authors, particularly in Grades 4 and 5. Problems of this pattern have sufficient practical importance to justify ample treatment on the part of authors. Further, there is no phraseology which is characteristic of the pattern as a whole; hence in solving problems of this pattern pupils must sense the underlying thought. This is a further justification for not neglecting this pattern. That it has been neglected in certain grades is evident from the graphic chart on page 33.

Comparisons and Relations Between Patterns

In the preceding discussion a study has been made of each of the ten subtraction patterns, particular emphasis being given to the characteristics of each pattern and the frequency of its treatment by the four different authors. A study will now be made of such relations as may exist between different patterns as disclosed by an analysis of the four texts.

Table I (page 43) gives data concerning the subtraction patterns when classified into four groups, these groups representing the four larger meanings of subtraction, namely, the remainder idea, the building-up idea, the comparison idea, and the separation-into-parts idea.

Attention is first called to the last line of Table I which gives, for each text, the *total number of repetitions* of the subtraction patterns in all four grades combined. These totals are 774 for

Text X, 796 for Y, 684 for Z, and 744 for W. Thus one finds surprising uniformity among the four authors in respect to their total treatment of subtraction, though a wide variation exists among them in their treatment of the individual patterns.

TABLE I

SUBTRACTION PATTERNS GROUPED ACCORDING TO THE FOUR IMPORTANT CONCEPTS OF SUBTRACTION

Pattern	*Text X*	*Text Y*	*Text Z*	*Text W*
S1 + S2 + S3 Remainder Idea	268 35%	262 33%	290 42%	341 46%
S4 Building-up Idea	71 9%	56 7%	84 12%	108 14%
S5 + S6 + S7 + S8 + S9 Comparison Idea	382 49%	415 52%	280 41%	250 34%
S10 Separation-into-Parts Idea	53 7%	63 8%	30 5%	45 6%
Total Offering	774	796	684	744

The number 268 above shows the total repetitions in all four grades of the three patterns in the remainder group. The 35% shows what per cent 268 is of 774 which represents the total offering in subtraction.

In the first line of Table I facts are given concerning the relative emphasis upon the remainder idea given by the four authors. Text Y devotes 33% of its total offering to the remainder idea; corresponding percentages for the other texts are 35% for X, 42% for Z, and 46% for W.

The third line of Table I is devoted to the comparison idea. Here one finds that two texts, X and Y, give practically half (49% and 52% respectively) of their total offerings to the comparison idea, which is considerably more attention than these two texts give to the remainder idea. Text Z gives about equal attention to these two ideas, namely, 41% to one and 42% to the other. Text W is the only one that gives more attention to remainder problems than to comparison problems; in fact, Text W practically reverses the policy followed by Texts X and Y.

TABLE II

SUMMARY OF THE TOTAL REPETITION OF EACH SUBTRACTION PATTERN IN ALL FOUR GRADES COMBINED

Text	*Item*	*S1*	*S2*	*S3*	*S4*	*S5*	*S6*	*S7*	*S8*	*S9*	*S10*	*Total*
X	Number of repetitions ...	185	66	17	71	139	64	89	87	3	53	774
	Per cent	23.9	8.5	2.2	9.2	18.0	8.3	11.5	11.2	0.4	6.8	100
	Rank	1	6	9	5	2	7	3	4	10	8	
Y	Number of repetitions ...	182	67	13	56	218	80	13	102	2	63	796
	Per cent	22.9	8.4	1.6	7.0	27.4	10.1	1.6	12.8	0.3	7.9	100
	Rank	2	5	8	7	1	4	8	3	10	6	
Z	Number of repetitions ...	198	51	41	84	83	41	13	138	5	30	684
	Per cent	28.9	7.5	6.0	12.3	12.1	6.0	1.9	20.2	0.7	4.4	100
	Rank	1	5	6	3	4	6	9	2	10	8	
W	Number of repetitions ...	225	92	24	108	91	59	23	72	5	45	744
	Per cent	30.2	12.4	3.2	14.5	12.2	7.9	3.1	9.7	0.7	6.1	100
	Rank	1	3	8	2	4	6	9	5	10	7	

Note: Table II shows for Text X that S1 has 185 repetitions in all four grades, which is 23.9% of the total subtraction offering of 774 repetitions.

The second and fourth lines of Table I give data concerning the building-up idea and the separation-into-parts idea. It is evident from the table that of the two ideas the building-up idea is given the greater prominence. These two ideas, of course, differ from the remainder idea. If, for each text, we combine the figures for these two ideas with the figures for the comparison idea, we find that for Texts X and Y practically two thirds of the entire offering in subtraction is devoted to these three ideas, leaving only one third of the offering for the remainder idea. Texts Z and W give a little more than one third of their offerings to the remainder idea but neither of these texts goes so far as to devote half of its offering to remainder problems. The point to be emphasized is that the remainder idea gets only one third the attention of two authors and less than half the attention of the other two.

It will be helpful at this point to review briefly the historical development of subtraction. In the early history of the race subtraction was used only in connection with the remainder idea, the question being, "How much is left?" It was with considerable difficulty that the race later on developed the comparison idea of subtraction. The history of the race in this respect is repeated in the schoolroom today since the comparison types of problems are more difficult for pupils than the take-away types. To compare two numbers by subtraction requires an abstract type of thinking. In view of the difficulty of the comparison concept, it is of interest that three out of the four texts give so much attention to that type of work.

The discussion above has been concerned with the arrangement of patterns in groups. Consider next such relations as may exist between the separate patterns. Table II, on page 44, shows for each text the total number of repetitions of each pattern in all four grades combined. In this table it is found that in three out of the four texts Pattern S1 has by far the greatest emphasis. This pattern represents 23.9% of the total offering in subtraction in Text X, 28.9% in Z, and 30.2% in W. In Text Y Pattern S1 holds second place, with 22.9% of the total repetitions. Pattern S1 is the most important pattern of the take-away group, with a wide diversity of situations in which it can be applied. Hence it is not surprising to find that it occupies first rank among the ten patterns in three out of the four texts. A further examination of Table II shows, at the other extreme, that Pattern S9 ranks lowest in all four texts. Between these extremes there are eight other patterns which vary considerably in their emphasis from text to text.

The "total repetitions" of each pattern, as given in Table II, are shown in graphic form in Chart 1 on page 34. This chart makes it possible to compare the treatment of the ten subtraction patterns in any one text with the corresponding treatment in each of the other texts.

Should Certain Patterns Be Combined?

Throughout this study ten different subtraction patterns have been discussed, though these patterns have been classified into four groups representing the four larger meanings of subtraction. The analysis of the four series of textbooks, however, has been in terms of the ten distinct patterns. Since certain patterns like S3 and S9 were found relatively few times compared with other patterns, the question might arise concerning the advisability of abandoning these as separate patterns and merging them with other patterns to which they are closely related. For example, since Pattern S3 is the inverse of Patterns S1 and S2, would it be better to merge certain problems of S3 with S1 and others with S2? While this could be done, certain advantages that come from retaining S3 would be lost. It must be recalled that S3 represents the *inverse cases* under S1 and S2 which, if merged with S1 and S2, would leave one without any indication as to the relative attention that was being paid to the inverse cases. While problems representing inverse cases do not arise in life as frequently as direct cases, inverse cases are, nevertheless, important. Consequently, it seems desirable to retain Pattern S3 to assure representation of these inverse cases. In like manner, S9 is represented by a very few problems. In fact, it is the pattern which is given the least attention by all four authors. Pattern S9 is the inverse of four comparison patterns. If S9 were distributed over these four comparison patterns, again one would lose evidence of the relative attention being paid to these inverses.

Another type of abridgment that might be suggested would be to unite Patterns S5 and S6. Pattern S5 compares numbers by using the phrases *how much more* and *how much less;* Pattern S6 uses such phrases as *how much larger, how much older, how much smaller.* It is evident that the two patterns are closely related, but if merged the relative attention given each would be lost. Hence there seem to be advantages in keeping these two patterns distinct as they are now.

For purposes of reference complete tables are given at the end of this book incorporating the results of the analysis of the four texts with respect to the ten subtraction patterns. These results appear in Tables XIVa, XIVb, XIVc, and XIVd in the Appendix. One of these tables corresponds to each of the four texts. These tables include the number of repetitions of each pattern found both in one-step and in multiple-step problems. In the case of the multiple-step problems, the tables give the repetitions of the patterns found both in the hidden steps and in the final steps of such problems. The graphic charts of each of the ten patterns which have appeared on the preceding pages of this chapter were based upon the four tables mentioned above. The graphic charts, however, do not differentiate between the hidden and the final steps. Instead, in the charts, the hidden and final steps are combined as a single item to form the "solid" part of each bar.

CHAPTER IV

DIVISION PATTERNS

Every division problem represents one of two fundamental meanings—measurement or partition. The measurement concept is illustrated by this problem: "In playing a game 30 boys were divided into groups, each group containing 6 boys. How many groups were there?" In this case 30 boys are being measured in terms of a unit group containing 6 boys and it is found that 30 boys contain this group 5 times. The partition concept is illustrated by this problem: "If 20 apples are divided equally among 5 boys, how many apples will each boy get?" It is found that each boy will get 4 apples. Thus it is seen that partition is concerned with finding the *size* of each group, whereas the measurement aspect of division is concerned with finding how many groups of a given size there will be. In problems of the *measurement* type, it is apparent that the dividend and the divisor are each concrete numbers while the quotient is abstract. In *partition* problems, the dividend and the quotient are each concrete while the divisor is abstract. In short, if the divisor is concrete, the problem is of the measurement type; if the divisor is abstract, the problem is of the partition type.

A critical examination of all the division problems found in the four series of textbooks reveals sixteen patterns, eight of which represent the measurement concept of division and eight the partition concept. Certain relationships found among the patterns which involve the measurement concept of division make it possible to classify these patterns in three major groups: (I) the simpler patterns requiring that a number of groups be formed, (II) the patterns involving concepts of distance, rate, and time, and (III) the patterns involving the comparison of two numbers. Certain relationships found among the patterns which involve the partition concept of division make it possible to also classify these patterns in three major groups: (IV) the patterns involving the finding of the amount, the size, or the cost of each part, (V) the pattern involving the concept of an average not related to distance, rate, and time, and (VI) the patterns involving concepts of an average in relation to distance, rate, and time. The distinctions between these six major groups of division patterns and between the patterns of a given group will become clear from the discussions and illustrations that follow.

Patterns Representing the Measurement Concept of Division

I. Simpler patterns requiring a number of groups to be formed

D1. Given a *total number* of like units and the *number of units in each group,* find the *number of groups* that can be formed.

(1) The children made 200 pieces of candy to sell at the school fair. They put 5 pieces in each bag. How many bags of candy did they have?

(2) There were 36 children at Betty's party. If 4 children sat at each table for lunch, how many tables were used?

D2. Given the *total cost* or *amount* and the *cost* or *amount of one unit,* find the *number of units* purchased or required. The word *amount,* as used in this pattern, refers wholly to money.

(1) Tom had 30¢ to spend for stamps. If one stamp costs 3¢, how many stamps can he get for 30¢?

(2) George gets 5¢ each time he runs an errand. He wants to buy a cap that costs 40¢. How many errands will he have to run to earn 40¢?

D3. Given the *total amount* to be *paid, saved,* or *done* and the *rate* of doing it, find the *total time* needed.

(1) Jack wants a pair of skates which he saw in a store window.

The price is $4.80. Jack can save $.80 a week. In how many weeks can he save enough money to buy the skates?

(2) The pupils in the fifth grade use 42 pencils a month. The teacher has 300 pencils on hand. How long will the 300 pencils last?

The underlying thought in each of the above three patterns is that of finding the *number of groups* that can be formed when the *total number of like units* and the *number of like units in one group* are given. In Pattern D1 the units to be divided into groups include everything except money, while in Pattern D2 the units to be divided into groups represent money only. In Pattern D3 the question is concerned with finding the length of time it will take to do a piece of work or to save a specified amount of money. The answer is always in terms of time—hours, days, weeks, months, or years.

The second group of patterns representing the measurement concept of division is made up of problems that are concerned with distance, rate, and time, frequently called motion problems.

II. The patterns involving concepts of distance, rate, and time

D4. Given the *total distance* and the *rate per unit of time*, find the *time*.

At the rate, average speed, averages so many miles an hour are technical expressions characteristic of this pattern.

(1) Traveling at 60 miles an hour, how long will it take a train to go 180 miles?

(2) If I drive at a rate of 25 miles an hour, how long should I allow for a trip of 123 miles?

(3) Jack's father averages 34 miles an hour in his new car. How many hours will it take him to make a trip of 408 miles to New York?

D5. Given the *total distance* and the *distance covered per gallon*, find the *number of gallons*.

Miles per gallon, at the rate, averages so many miles on

DIVISION PATTERNS

DIVISION PATTERN D1

Given the Total Number and the Number in Each Group, to Find the Number of Groups

☐ Repetitions in one-step problems

■ Repetitions in multiple-step problems

Grade 3

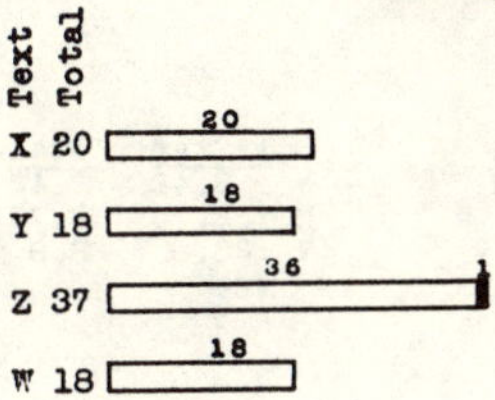

	1-Step	Total
σ	7.55	7.98
M	23.00	23.25
V	32.8	34.3

Grade 4

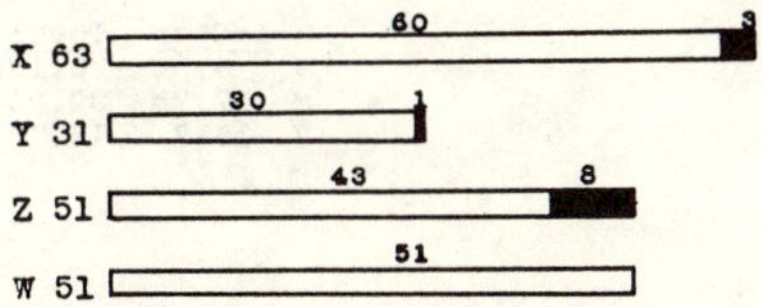

	1-Step	Total
σ	11.02	11.49
M	46.00	49.00
V	24.0	23.4

Grade 5

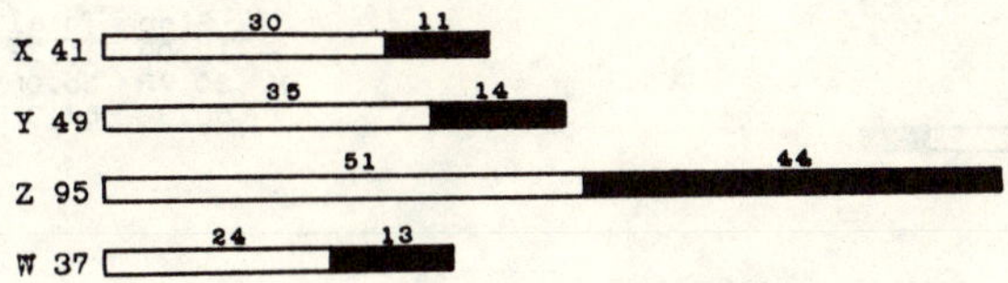

	1-Step	Total
σ	10.02	23.21
M	35.00	55.50
V	28.6	41.8

Grade 6

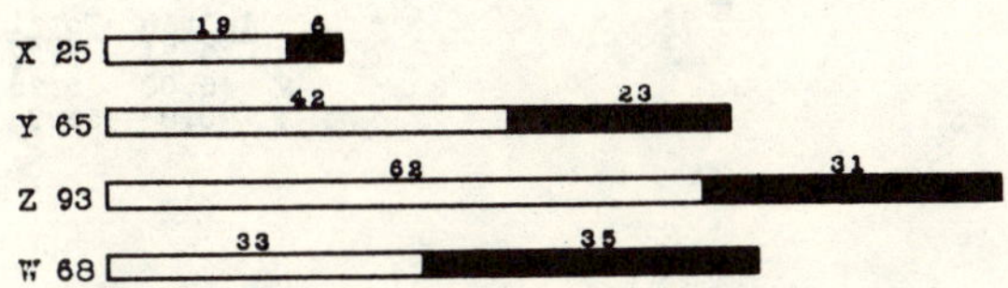

	1-Step	Total
σ	15.60	24.36
M	39.00	62.75
V	40.0	38.8

Grades 3 to 6 Combined*

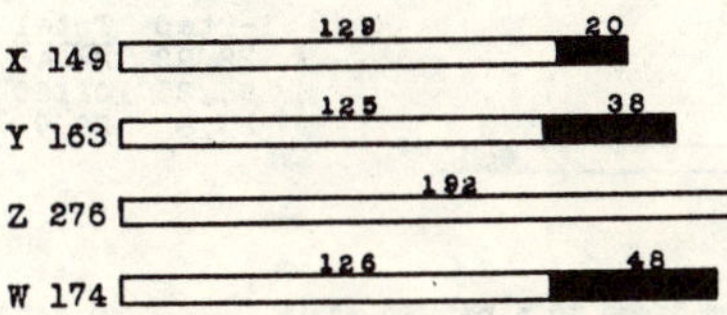

	1-Step	Total
σ	28.33	50.15
M	143.00	190.50
V	19.8	26.3

*The graph for Grades 3 to 6 Combined has been drawn to a reduced scale.

DIVISION PATTERN D2

Given the Total Cost and the Cost of One Unit, to Find the Number of Units Purchased

Grade 3

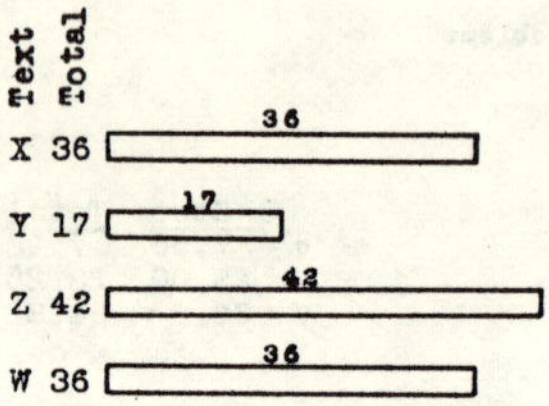

	1-Step	Total
σ	9.42	9.42
M	32.75	32.75
V	28.8	28.8

Grade 4

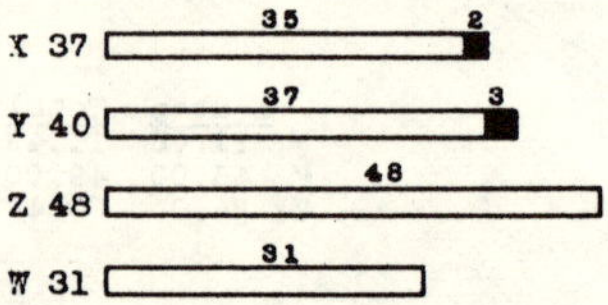

	1-Step	Total
σ	6.30	6.12
M	37.75	39.00
V	16.7	15.7

Grade 5

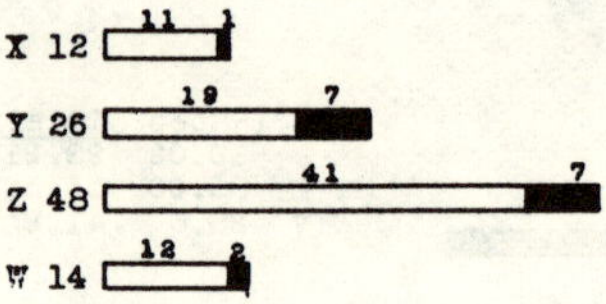

	1-Step	Total
σ	12.09	14.32
M	20.75	25.00
V	58.3	57.3

Grade 6

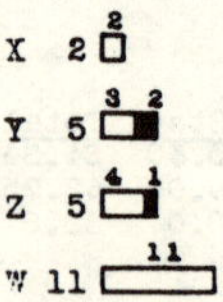

	1-Step	Total
σ	3.54	3.27
M	5.00	5.75
V	70.8	56.9

Grades 3 to 6 Combined*

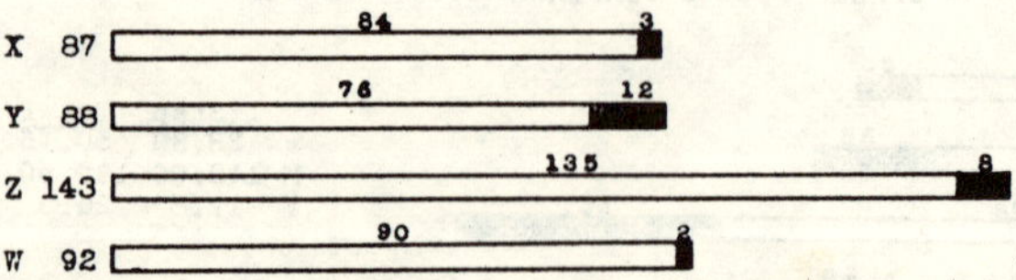

	1-Step	Total
σ	22.92	23.46
M	96.25	102.50
V	23.8	22.9

*The graph for Grades 3 to 6 Combined has been drawn to a reduced scale.

DIVISION PATTERN D3

Given the Total Amount Paid or Saved
and the Rate of Doing It,
to Find the Total Time Needed

Grade 3

Text	Total	Bar segments
X	8	8
Y	1	1
Z	2	2
W	5	5

V
1-Step 68.5
Total 68.5

Grade 4

Text	Total	Bar segments
X	2	2
Y	2	2
Z	4	3 1
W	8	8

V
1-Step 66.4
Total 61.3

Grade 5

Text	Total	Bar segments
X	6	4 2
Y	10	7 3
Z	13	8 5
W	0	

V
1-Step 65.5
Total 67.2

Grade 6

Text	Total	Bar segments
X	3	2 1
Y	8	7 1
Z	10	9 1
W	10	7 3

V
1-Step 41.4
Total 36.9

Grades 3 to 6 Combined

Text	Total	Bar segments
X	19	16 3
Y	21	17 4
Z	29	22 7
W	23	20 3

V
1-Step 12.7
Total 16.3

DIVISION PATTERN D4

Given the Total Distance and the Rate,
to Find the Time

Grade 3

Text	Total	Bar segments
X	0	
Y	0	
Z	0	
W	0	

Grade 4

Text	Total	Bar segments
X	1	1
Y	0	
Z	9	9
W	1	1

V
1-Step 132.0
Total 132.0

Grade 5

Text	Total	Bar segments
X	8	7 1
Y	9	9
Z	19	19
W	5	5

V
1-Step 53.9
Total 51.3

Grade 6

Text	Total	Bar segments
X	8	7 1
Y	7	5 2
Z	6	5 1
W	0	

V
1-Step 60.9
Total 59.2

Grades 3 to 6 Combined

Text	Total	Bar segments
X	17	15 2
Y	16	14 2
Z	34	33 1
W	6	6

V
1-Step 58.1
Total 55.1

DIVISION PATTERN D5

Given the Total Distance
and the Distance Covered per Gallon,
to Find the Number of Gallons

Grade 3

Text	Total
X	0
Y	0
Z	0
W	0

Grade 4

Text	Total	Bar
X	0	
Y	1	1
Z	0	
W	4	4

Grade 5

Text	Total	Bar
X	0	
Y	1	1
Z	2	1 (open), 1 (shaded)
W	0	

Grade 6

Text	Total	Bar
X	2	2
Y	1	1
Z	3	2 (open), 1 (shaded)
W	0	

Grades 3 to 6 Combined

Text	Total	Bar
X	2	2
Y	3	3
Z	5	3 (open), 2 (shaded)
W	4	4

DIVISION PATTERN D6

Given the Total Time
and the Time per Unit of Distance,
to Find the Total Distance

Grade 3

Text	Total
X	0
Y	0
Z	0
W	0

Grade 4

Text	Total	Bar
X	0	
Y	0	
Z	1	1
W	0	

Grade 5

Text	Total	Bar
X	0	
Y	0	
Z	1	1
W	0	

Grade 6

Text	Total	Bar
X	1	1
Y	1	1
Z	1	1
W	0	

Grades 3 to 6 Combined

Text	Total	Bar
X	1	1
Y	1	1
Z	3	3
W	0	

DIVISION PATTERN D7

Comparing Two Numbers to Find How Many Times as Large One Is As the Other

Grade 3

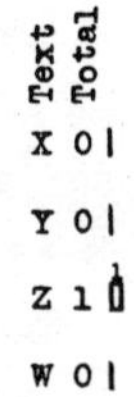

Grade 4

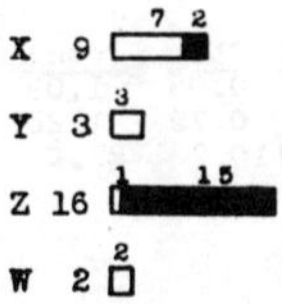

	1-Step	Total
σ	2.28	5.59
M	3.25	7.50
V	70.2	74.5

Grade 5

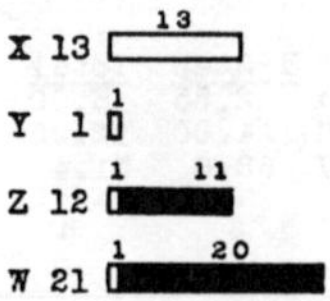

	1-Step	Total
σ	5.20	7.12
M	4.00	11.75
V	130.0	60.6

Grade 6

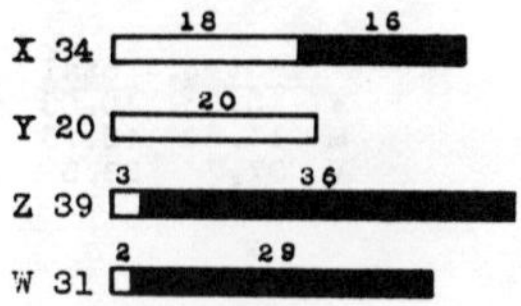

	1-Step	Total
σ	8.29	6.96
M	10.75	31.00
V	77.1	22.5

Grades 3 to 6 Combined

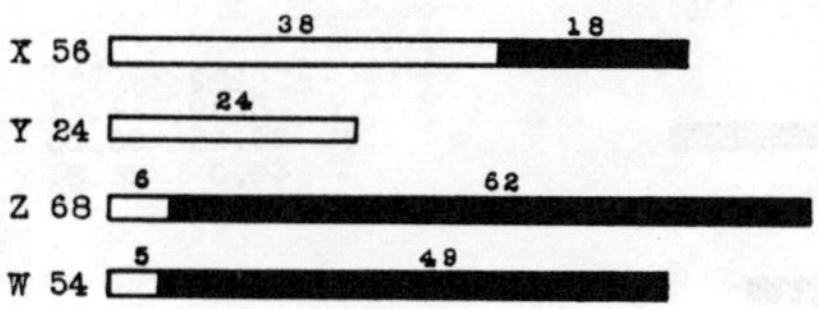

	1-Step	Total
σ	13.68	16.21
M	18.25	50.50
V	75.0	32.1

DIVISION PATTERN D8

Comparing Two Numbers to Find What Part the Smaller Number Is of the Larger

Grade 3

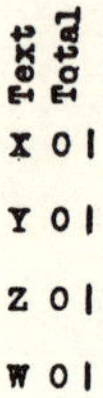

Grade 4

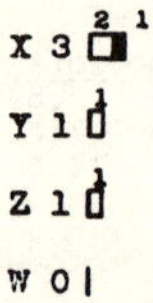

	1-Step	Total
σ	0.83	1.09
M	0.75	1.25
V	110.7	87.2

Grade 5

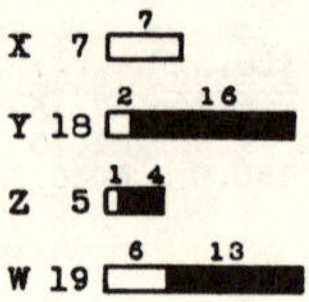

	1-Step	Total
σ	2.55	6.30
M	4.00	12.25
V	63.8	51.4

Grade 6

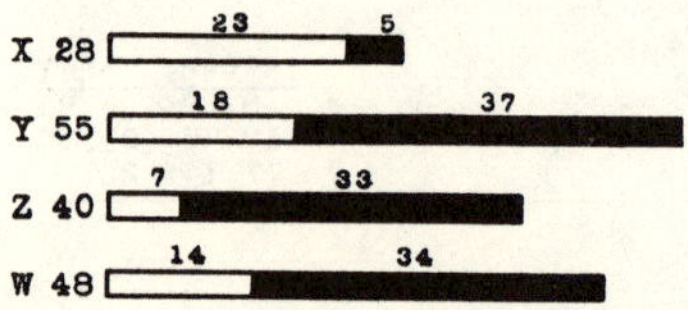

	1-Step	Total
σ	5.85	10.03
M	15.50	42.75
V	37.7	23.5

Grades 3 to 6 Combined

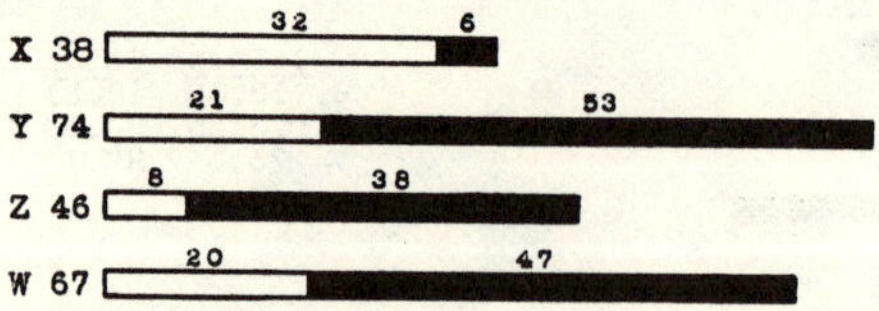

	1-Step	Total
σ	8.50	14.74
M	20.25	56.25
V	42.0	26.2

DIVISION PATTERN D9

Given the Total Number, to Find a Unit Fractional Part of It

Grade 3

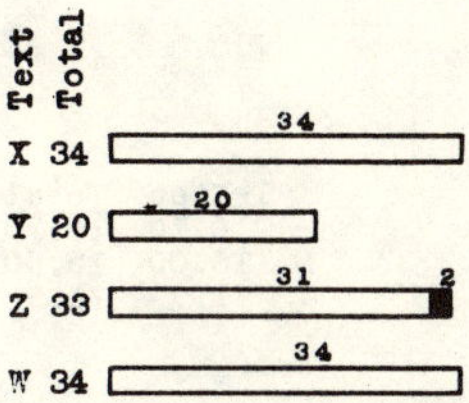

	1-Step	Total
σ	5.76	5.93
M	29.75	30.25
V	19.4	19.6

Grade 4

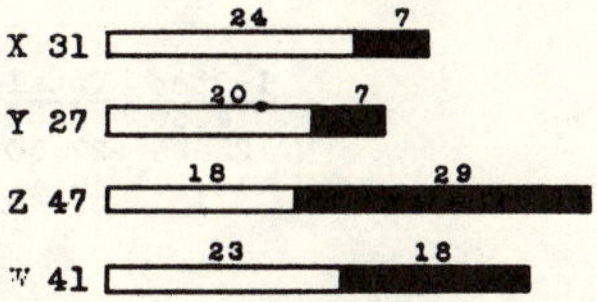

	1-Step	Total
σ	2.38	7.92
M	21.25	36.50
V	11.2	21.7

Grade 5

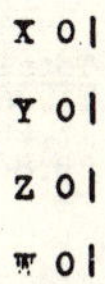

Grade 6

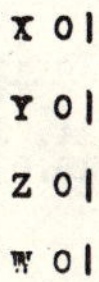

Grades 3 to 6 Combined

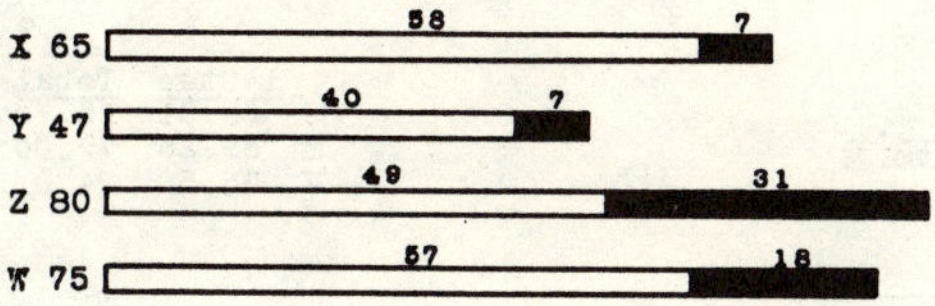

	1-Step	Total
σ	7.25	12.62
M	51.00	66.75
V	14.2	18.9

DIVISION PATTERN D10

Given a Total Amount and the Number of Equal Parts Into Which It Is to Be Divided, to Find the Size per Part

Grade 3

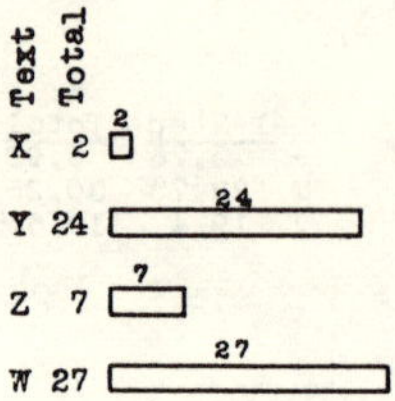

	1-Step	Total
σ	10.70	10.70
M	15.00	15.00
V	71.3	71.3

Grade 4

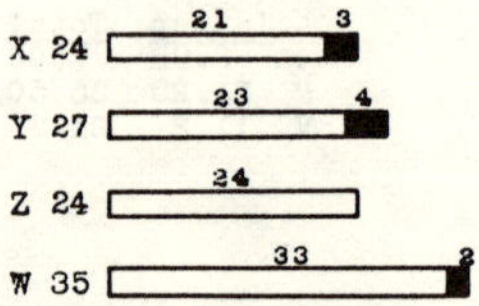

	1-Step	Total
σ	4.60	4.50
M	25.25	27.50
V	18.2	16.4

Grade 5

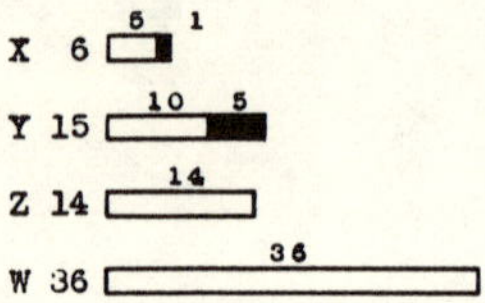

	1-Step	Total
σ	11.84	11.86
M	16.25	17.25
V	72.9	68.8

Grade 6

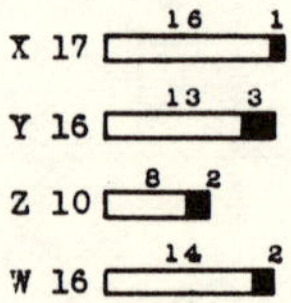

	1-Step	Total
σ	2.95	2.77
M	12.75	14.75
V	23.1	18.8

Grades 3 to 6 Combined*

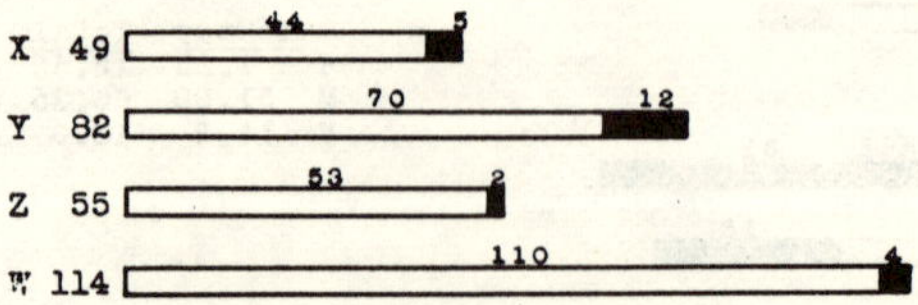

	1-Step	Total
σ	25.31	25.72
M	69.25	75.00
V	36.5	34.3

*The graph for Grades 3 to 6 Combined has been drawn to a reduced scale.

DIVISION PATTERN D11

Given a Total Amount and the Number of Persons, to Find the Share of Each Person

Grade 3

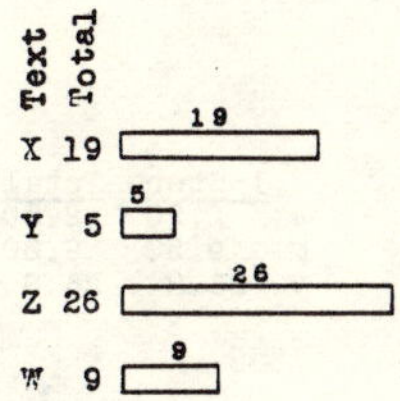

	1-Step	Total
σ	8.26	8.26
M	14.75	14.75
V	56.0	56.0

Grade 4

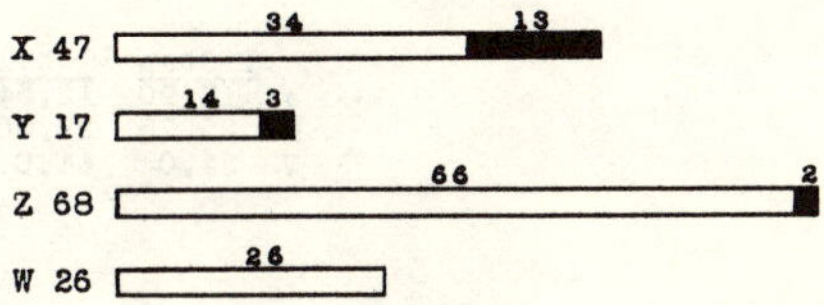

	1-Step	Total
σ	19.26	19.73
M	35.00	39.50
V	55.0	49.9

Grade 5

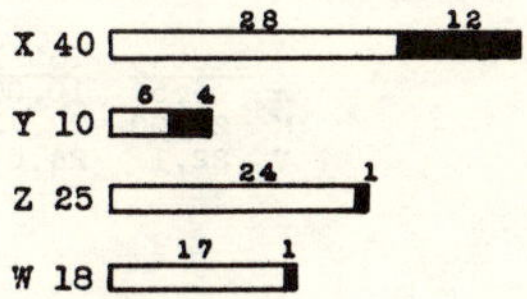

	1-Step	Total
σ	8.35	11.03
M	18.75	23.25
V	44.5	44.9

Grade 6

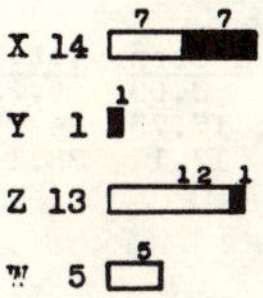

	1-Step	Total
σ	4.30	5.45
M	6.00	8.25
V	71.7	66.1

Grades 3 to 6 Combined*

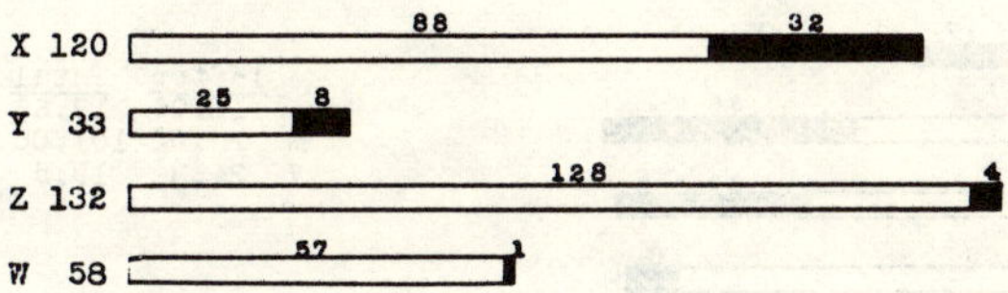

	1-Step	Total
σ	38.08	41.43
M	74.50	85.75
V	51.1	48.3

*The graph for Grades 3 to 6 Combined has been drawn to a reduced scale.

DIVISION PATTERN D12

Given the Total Cost and the Number of Like Articles Purchased, to Find the Cost of One Article

Grade 3

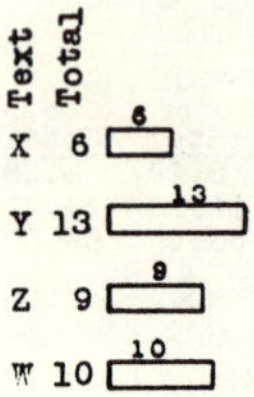

	1-Step	Total
σ	2.50	2.50
M	9.50	9.50
V	26.3	26.3

Grade 4

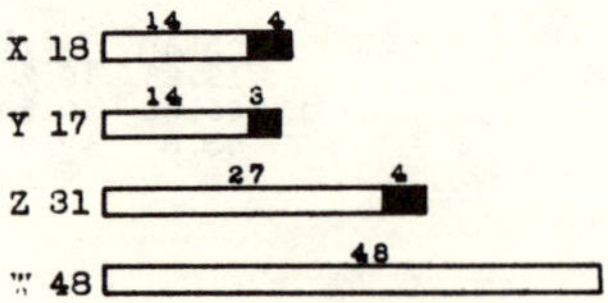

	1-Step	Total
σ	13.90	12.54
M	25.75	28.50
V	54.0	44.0

Grade 5

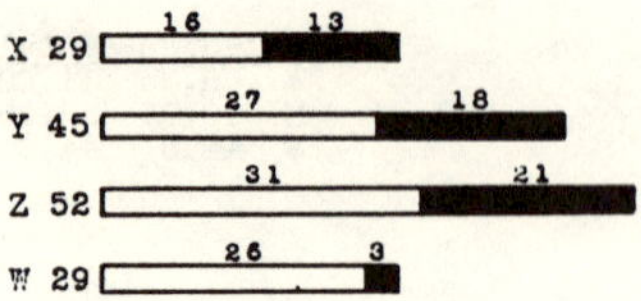

	1-Step	Total
σ	5.52	10.06
M	25.00	38.75
V	22.1	26.0

Grade 6

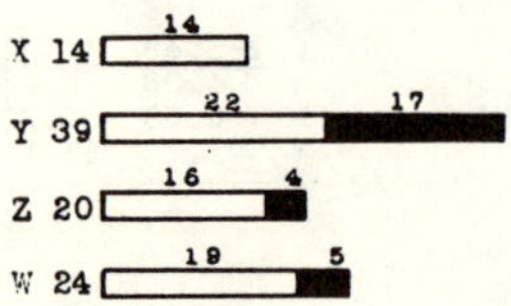

	1-Step	Total
σ	3.03	9.23
M	17.75	24.25
V	17.1	38.1

Grades 3 to 6 Combined*

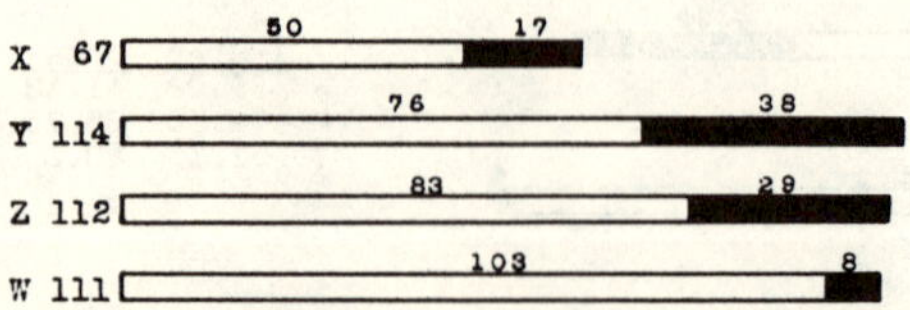

	1-Step	Total
σ	18.96	19.66
M	78.00	101.00
V	24.3	19.5

*The graph for Grades 3 to 6 Combined has been drawn to a reduced scale.

DIVISION PATTERN D13

Given the Total Amount and the Number of Units Involved, to Find the Average per Unit

Grade 3

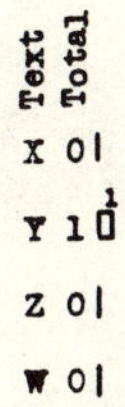

Grade 4

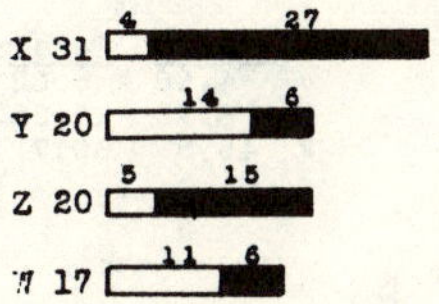

	1-Step	Total
σ	4.15	5.34
M	8.50	22.00
V	48.8	24.3

Grade 5

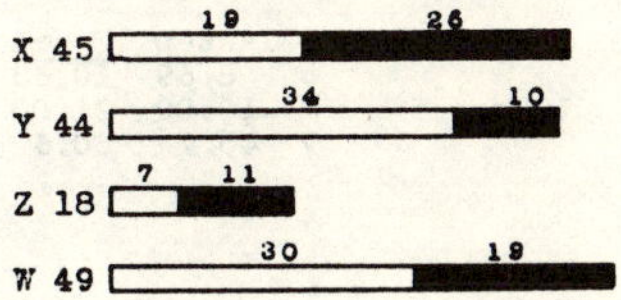

	1-Step	Total
σ	10.50	12.27
M	22.50	39.00
V	46.7	31.5

Grade 6

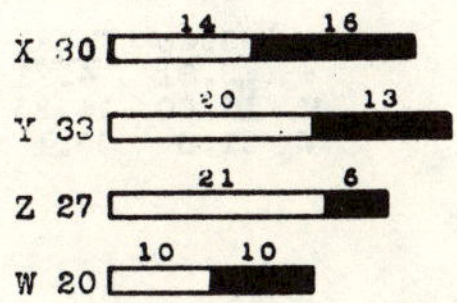

	1-Step	Total
σ	4.49	4.82
M	16.25	27.50
V	27.6	17.5

Grades 3 to 6 Combined*

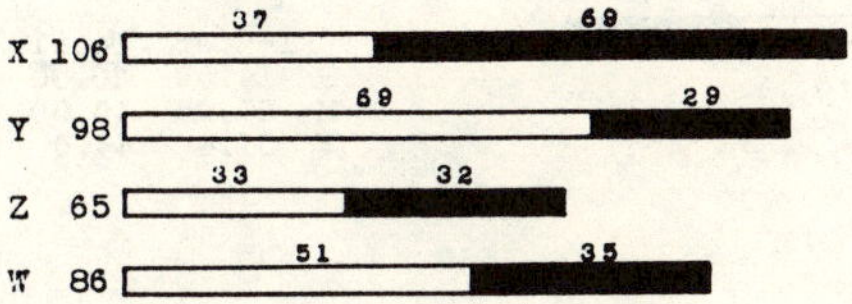

	1-Step	Total
σ	14.10	15.45
M	47.50	88.75
V	29.7	17.4

*The graph for Grades 3 to 6 Combined has been drawn to a reduced scale.

DIVISION PATTERN D14

Given the Total Distance and the Time, to Find the Rate

Grade 3

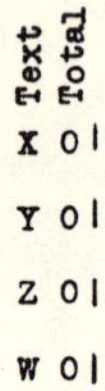

Grade 4

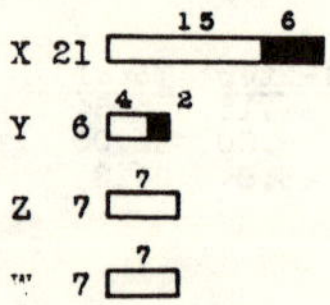

	1-Step	Total
σ	4.08	6.22
M	8.25	10.25
V	49.5	60.7

Grade 5

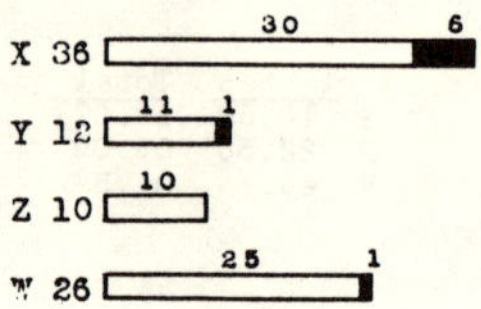

	1-Step	Total
σ	8.69	10.63
M	19.00	21.00
V	45.7	50.6

Grade 6

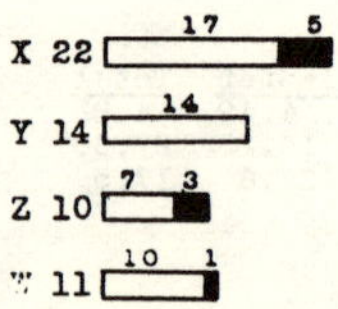

	1-Step	Total
σ	3.81	4.71
M	12.00	14.25
V	31.8	33.1

Grades 3 to 6 Combined

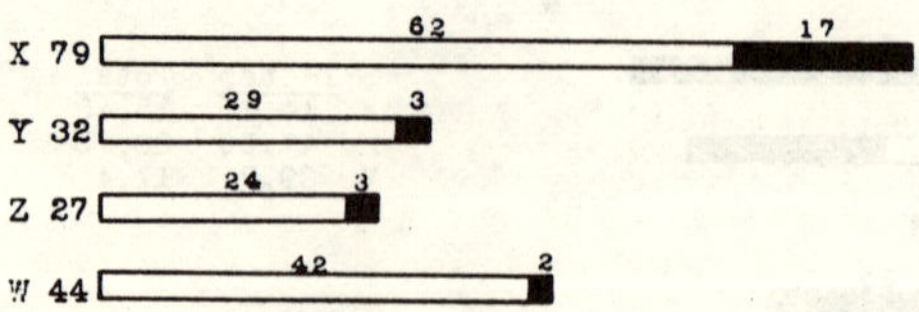

	1-Step	Total
σ	14.69	20.30
M	39.25	45.50
V	37.4	44.6

DIVISION PATTERN D15

Given the Total Distance
and the Number of Gallons,
to Find the Distance per Gallon

Text	Total	Bar
Grade 3		
X	0	
Y	0	
Z	0	
W	0	
Grade 4		
X	0	
Y	0	
Z	1	1
W	0	
Grade 5		
X	7	7
Y	1	1
Z	2	2
W	2	2
Grade 6		
X	4	3 1
Y	4	2 2
Z	5	2 3
W	0	
Grades 3 to 6 Combined		
X	11	10 1
Y	5	3 2
Z	8	5 3
W	2	2

DIVISION PATTERN D16

Given the Total Time and
the Total Number of Units
of Distance Covered,
to Find the Time to Cover One Unit

Text	Total	Bar
Grade 3		
X	1	1
Y	0	
Z	0	
W	1	1
Grade 4		
X	0	
Y	1	1
Z	0	
W	1	1
Grade 5		
X	0	
Y	2	1 1
Z	0	
W	5	5
Grade 6		
X	0	
Y	0	
Z	0	
W	2	2
Grades 3 to 6 Combined		
X	1	1
Y	3	2 1
Z	0	
W	9	7 2

CHART 2

Total Repetitions of Each of the Division Patterns
in All Four Grades Combined,
Arranged by Texts

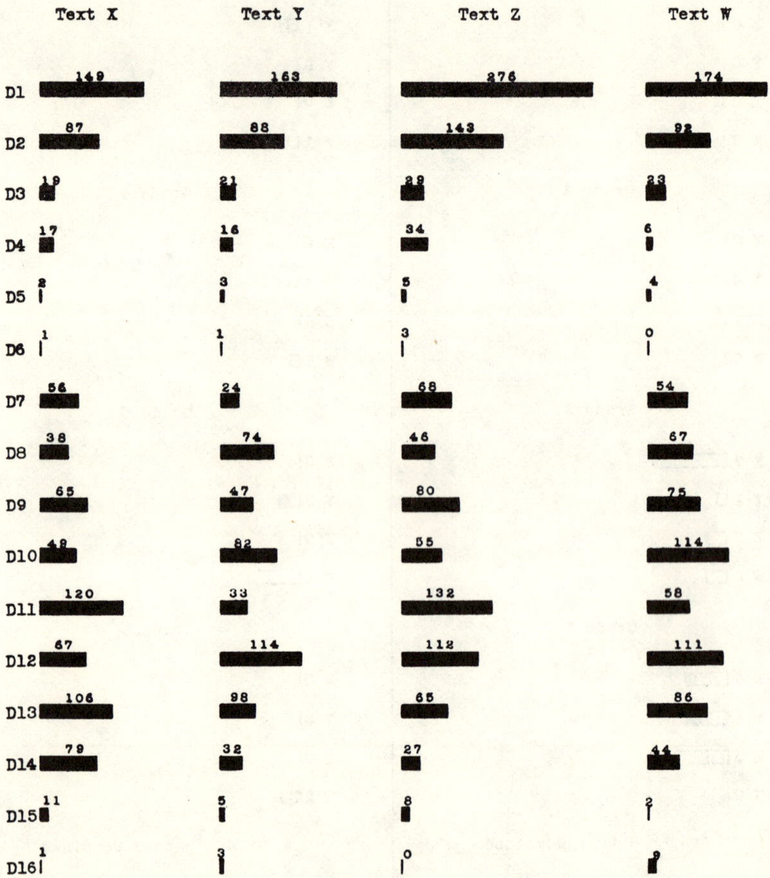

1 gallon are technical expressions characteristic of this pattern.

(1) Mrs. Brown's car uses a gallon of gasoline each 21 miles it travels. How many gallons will the car use on a trip of 672 miles?

(2) Jack's car averages 18.3 miles to a gallon of gasoline. On Tuesday the speedometer showed he had gone 91.5 miles. How many gallons of gasoline were used on Tuesday?

D6. Given the *total time* and the *time per unit of distance,* find the *total distance.* (This pattern is of rare occurrence.)

(1) If a fast mail train travels a mile in 1.25 minutes, how many miles does it travel in 31.25 minutes?

(2) Joseph rode his pony a mile in 15 minutes. At that rate, how far would he ride in 30 minutes?

(3) If it takes a quarter of an hour for John to walk a mile, how many miles could he walk in 3 hr.?

The underlying thought in the problems of each pattern is that of measuring, that is, finding how many times the distance traveled in one day or one hour is contained in the total distance, as in D4; or how many times the distance traveled on one gallon is contained in the total distance, as in D5; or how many times the time to travel a mile is contained in the total time traveled, as in D6. The expressions *at the rate, average, average speed, miles per gallon, averages so many miles an hour,* etc., are very technical and are characteristic of problems of these patterns.

The third group of patterns representing the measurement idea of division involves the comparison of two numbers. The comparison may be made in either of two ways, as shown in the following two patterns.

III. The patterns involving the comparison of two numbers

D7. Given *two numbers,* how *many times as large* is *one number as the other?*

Such expressions as *how many times as large, how many*

times as many, how many times as high, etc., are characteristic of this pattern.

(1) Daniel is 7 years old. His mother is 35 years old. How many times Daniel's age is his mother's age?
(2) Bob earned $9 this month. His big brother Bill earned $63. Bill earned ..... times as much as Bob.

D8. Given *two numbers,* the *smaller number* is *what part of the larger number?*

(1) The Reds won 21 games out of 24 baseball games. What part of the games the Reds played did they win?
(2) Toby had $18 and spent $12. What part of his money did he spend?

The problems in each of the eight patterns discussed above are alike in one respect, viz., they involve the division of one number by another number of the same kind, that is, finding how many groups of a given size can be formed from the given number. This is the *measurement concept* of division. The child arrives at this generalization only after he has solved many problems under each pattern. As the problems of each pattern contribute to the generalization about that pattern, so each pattern contributes to the formulation of the larger generalization regarding the measurement concept of division. Though the social situations and the language in which these problems are expressed differ widely, the underlying thought is the same —finding how many groups of a given size can be formed from the given number.

The problems that contribute to the *partition concept* of division have been classified in three major groups of patterns, the discussion of which follows.

Patterns Representing the Partition Concept of Division

IV. THE PATTERNS THAT REQUIRE THE FINDING OF THE AMOUNT, THE SIZE, OR THE COST OF EACH PART

D9. Given the *total number,* find a *fractional part* of it (the numerator of the fraction being one). Pattern D9 is a division pattern for Grades 3 and 4.

(1) Bob has 10 apples. He gives half of them to Peggy. How many apples does he gives Peggy?
(2) Mary and Tom bought a bag of popcorn together for 12¢. Each paid ½ of the cost. How much did each one pay?
(3) Billy puts ⅓ of all the money he earns in the bank. He earned 69¢ last week. How much did he put in the bank?

D10. Given the *total amount* and the *number of parts* into which it is to be equally divided, find the *number* or *size* per part.

(1) I want to plant 8 rows of tomato plants in my garden. How many plants should I put in each row if I have 72 plants and if I put the same number of plants in each row?
(2) Five busses took 175 children to a picnic. If each bus took the same number of children, how many children rode in each bus?
(3) Alice has a piece of blue ribbon 20 in. long. She wants to cut it into 4 pieces so that all the pieces will be of the same length. How many inches long will each piece be?

D11. Given the *total amount* and the *number* of persons involved, find the equal *share* of each person. (This is the sharing concept.)
Divide equally and *share equally* are expressions characteristic of this pattern.

(1) Uncle Bob has 235 stamps. He *divides* them *equally* among 5 boys. How many stamps does each boy get as his share?
(2) If 5 children buy a croquet set for $2.75 and *share* the cost *equally,* how much should each child pay?

D12. Given the *total cost* and the *number* of like articles purchased, find the *cost of one* article.

(1) Mae paid 18 cents for 3 grapefruit. How much did they cost apiece?
(2) Mrs. Day paid 39¢ for 3 yards of ribbon. What was the cost of 1 yard?
(3) Harry paid 20 cents for 4 bananas. Find the price of one banana.

Each problem of these four patterns requires finding the amount, the size, or the cost of each part when the total number and the number of parts are given. This is the partition meaning of division. In Pattern D9 the problems are so stated that the fractional part of the number to be found is *actually expressed* by giving the fractions in symbolic form, such as $\frac{1}{2}$ or $\frac{1}{3}$, or by stating the fractions in verbal form, such as *one-half* or *a third of*. The child is first confronted with this type of problem soon after he begins to study division; he is then taught that $\frac{1}{2}$, $\frac{1}{3}$, $\frac{1}{4}$, etc., of a number can be found by dividing the number by 2, 3, 4, etc. At this early stage these problems are thought of as belonging to division. Later on, however, when the pupil is in Grade 5, a problem like $\frac{1}{4}$ of 20¢ is considered a multiplication problem and is usually written $\frac{1}{4} \times$ 20¢. To think of $\frac{1}{4}$ of 20¢ as a multiplication problem rather than a division problem involves a new concept of multiplication, which is usually taught in Grade 5.

In Patterns D10, D11, and D12, which are a part of this group, a fractional part of a number is to be found, but the fractional part to be found is not expressed as a fraction, as it is in the problems of Pattern D9. For example, in Problem (1) of Pattern D10 (given above) one wishes to know the number of plants to be put in each row if there are 72 plants and there are to be 8 rows. In this case the number of parts into which 72 is to be divided is expressed by the number 8. If this had been a problem of Pattern D9, the problem would have contained the fraction $\frac{1}{8}$ rather than 8.

The problems of Pattern D11 involve the sharing concept. Such expressions as *divide equally, share equally, how much is each one's share* are characteristic of this pattern and help emphasize the basic thought, which is that of sharing or dividing equally among several persons. The element common to all problems of Pattern D12 is that of finding the *cost of one* article when the *total cost* and *number* of like articles purchased are known.

The next group, representing the partition concept of division, consists of only one pattern, which is given below.

V. THE PATTERN INVOLVING THE CONCEPT OF AN AVERAGE. (AVERAGE PROBLEMS INVOLVING THE CONCEPT OF DISTANCE ARE NOT TO BE INCLUDED UNDER THIS PATTERN SINCE SUCH PROBLEMS ARE TO BE CLASSIFIED UNDER D14, D15, OR D16.)

D13. Given the *total amount* and the *number of units* involved, find the *average* per unit.

Such technical expressions as *average amount, average price, average cost, average rainfall, average yield,* etc., are generally characteristic of this pattern. Certain problems (few in number) are classified under this pattern where an *average* is implied but the word *average* is not used; see Problem (4) below.

(1) Our milk bill for December was $5.58. This was an *average* of ..... cents a day.

(2) Mr. Smith's orchard contained 60 apple trees. He raised 480 baskets of apples. What was the *average yield* a tree?

(3) Mary gained 34.5 lb. in 3 yr. What was the *average number* of pounds she gained per year?

(4) Mary worked 8 examples in 32 seconds. At this rate, how many seconds did Mary take to work one example?

The basic idea in each of the problems of this pattern is that of finding an average. This is a difficult concept for children. To complicate matters further, such technical expressions as *average amount, average price, average cost, average rainfall, average yield,* etc., are used. To be sure, these expressions help to emphasize the underlying thought but they add to the difficulty of the pattern.

The next group of patterns representing the partition concept is also concerned with finding an average and is probably more difficult for many children than Pattern D13.

VI. THE PATTERNS INVOLVING CONCEPTS OF AVERAGE IN RELATION TO DISTANCE, RATE, AND TIME

D14. Given the *total distance* and the *time,* find the *rate* per unit of time.

Such technical expressions as *miles per hour, average*

speed per hour, average rate, etc., are generally characteristic of this pattern. In some problems an average is implied but the word *average* is not used.

(1) Mr. Hill drove his car 288 miles in 8 hours. How many miles did he drive in 1 hour?

(2) Mr. Brown drove his car 775 miles in 25 hours. *How many miles per hour* did he *average?*

(3) An airplane went 1233 mi. in 9 hr. Find the *average speed* of the plane *per hour.*

D15. Given the *total distance* and the *number of gallons,* find the *distance* per gallon.

Such technical expressions as *miles per gallon, average number of miles per gallon, average per gallon* are characteristic of this pattern.

(1) Mr. Burns drove 544 miles, using 34 gallons of gasoline. How many miles did the car go on one gallon of gasoline?

(2) Mr. Lee used 23 gallons of gasoline on a trip of 506 miles. How *many miles per gallon* did he *average?*

D16. Given the *total time* and the *total distance* covered, find the *time* required to cover *one unit of distance.*

(1) Jack went on a 4-mile hike. He walked the whole distance in 96 minutes. How long did it take him, on the average, to walk a mile?

(2) An automobile made a 20-mile trip in 30 minutes. How many minutes did it take the automobile to go a mile?

Patterns D14, D15, and D16 are perhaps the most difficult patterns related to the partition concept of division, for, in these three patterns, the concept of an average is combined with the ideas of distance, rate, and time, which present difficulties in and of themselves. In the problems of Pattern D14, such expressions as *miles per hour, average speed per hour, average rate per hour,* etc., occur frequently. They are very technical and very difficult for some children. Yet an understanding of these expressions is necessary for the successful solution of these problems. These expressions help emphasize the underlying thought of the problems, which is that of finding the average rate to be traveled in one hour in order to cover the total distance in the given time.

The problems of D15 require the average distance traveled on one gallon of gasoline when the total distance traveled and the number of gallons used are known. Here again the technical expressions *miles per gallon, average number of miles per gallon,* etc., help emphasize the underlying thought, which is that of finding the average distance traveled on one gallon of gasoline.

It will be recalled that Pattern D15 refers specifically to *gallons of gasoline,* which makes it the most restricted of all the patterns in its application. The reason for this restriction is that the 15,000 problems examined in this study yielded very few problems of this pattern, all of which were in terms of *gallons of gasoline.* There is no reason why this pattern should be so restricted. The following problem would naturally come under this pattern if any such had been found, but this was not the case: "A railroad engine burns 5 tons of coal on a trip of 185 miles. How many *miles per ton* does it average?" It must be admitted, however, that this problem would be of interest mainly to engineers and that it does not begin to have the wide social application to the average individual that is possessed by problems of the gasoline type. No doubt this explains why the textbooks have limited their offering of problems of this kind to those dealing with the relation of gasoline consumption to distance. Attention is called at this time to Pattern D5, which is similarly limited to situations involving gasoline consumption and distance. Pattern D5 is, in fact, the counterpart of D15; D5 represents the *measurement* concept of division while D15 represents the *partition* concept.

Pattern D16 is very similar to Pattern D14 in that it involves the concept of an average in relation to distance and time. In Pattern D14, however, the *average distance traveled in one hour* is required, while in Pattern D16, the *average time to travel one mile* is required.

It will be noted that all the problems illustrating the eight patterns in the partition group are alike in that they involve the division or separation of a number into equal parts. Separation into equal parts may require finding the number in or the size of each part, finding the cost of one, finding the share of each person, finding the average cost, yield, or amount per unit, finding the rate traveled per unit of time, finding the distance traveled per gallon, or finding the time to travel one unit of distance. In each case, the question is concerned with the *number* or the *amount* in each part, and this is found by division.

As each problem belonging to a specific pattern contributes to the formulation of a generalization about that pattern, so each pattern contributes to the formulation of certain larger generalizations regarding the process of division. Finally, division comes to mean the process which is used (*a*) to find the number of groups of a given size that can be formed from a given number, (*b*) to find the number in each part, given the total number and the number of equal parts into which it is to be divided. These broad generalizations regarding division are reached only after wide experience with the problems of each pattern and recognition of the common elements. Understanding of these broad generalizations is very important for success in problem solving.

Distribution of the Sixteen Division Patterns by Grades in Four Series of Arithmetics

A critical study of the findings on division as presented in the four series of textbooks that were examined will show how the four authors have treated each of the division patterns and what emphasis they have given to the two broad concepts of division. First to be considered are Patterns D1, D2, and D3, which are the easiest patterns involving the measurement concept of division.

Division Pattern D1

The graphic chart of Pattern D1 is shown on page 53. On

this chart the graph for Grade 3 shows close agreement in Texts X, Y, and W with reference to the total number of repetitions of Pattern D1. In Text Z the "total repetitions" are about twice as many as in each of the other three texts. One striking feature in the graph for Grade 3 is that all repetitions of Pattern D1, save one in Text Z, occur in one-step problems.

In the graph for Grade 4 there is wide variation in the "total repetitions" of Pattern D1, these totals varying from 31 to 63. By far the greater number of appearances of this pattern occur in one-step problems. In Text W, D1 occurs exclusively in one-step problems.

The graph for Grade 5 shows fairly close agreement among Texts X, Y, and W on the "total repetitions" of this pattern; Text Z has more than twice as many as each of the other three texts. Pattern D1 still appears much more frequently in one-step problems than in multiple-step problems. In Text Z, however, there is a generous offering in multiple-step problems. In Grade 6 the variation among the texts in "one-step repetitions" and in "total repetitions" is more pronounced than in any of the other three grades. Repetitions in the one-step problems vary from 19 to 62, in the multiple-step problems from 6 to 35, while the "total repetitions" vary from 25 to 93.

The graph for all four grades combined shows close agreement among Texts X, Y, and W with reference to their "one-step repetitions" of Pattern D1 and fairly close agreement in their "total repetitions." Text Z has the largest offering of this pattern, which is due to the great emphasis on this pattern by Text Z in Grades 5 and 6.

Division Pattern D2

Pattern D2, like Pattern D1, belongs to the group of three patterns representing the simpler patterns of the measurement idea of division. Both patterns are closely related, but Pattern D1 covers a much wider range of situations than Pattern D2. The latter is concerned with one type of situation, namely, that of finding how many items can be purchased with a certain sum of money when the price of one is known.

The graphic chart of Pattern D2, given on page 54, rather strikingly presents the following facts:

In all four grades Pattern D2 appears almost exclusively in the form of "one-step repetitions"; that is, repetitions in multiple-step problems are negligible.

In Grade 5 there is a wide variation in the offerings; for example, Text Z gives 48 "total repetitions" as against 12 such repetitions by Text X. In Grade 6 there is a marked falling off in each text in the treatment of this pattern. It is odd that Text Z, which had 48 "total repetitions" in Grade 5 should drop to 5 repetitions in Grade 6. The meager offering of Pattern D2 in Grade 6 is rather significant since this is the grade in which division of a decimal by a decimal is normally taught, this topic giving excellent opportunities for practice in dividing dollars and cents by dollars and cents.

In all four grades combined the predominance of "one-step repetitions" is still in evidence. The total offering of Text Z is about one and one half times as large as that of each of the other three texts. If the combined graphs for all four grades of Patterns D1 and D2 are compared, it will be seen that Pattern D1 receives much greater emphasis than D2 in all four series. Texts Y, Z, and W, for example, give almost twice as much emphasis to D1 as to D2.

Division Pattern D3

From the graphic chart of Pattern D3, given on page 55, it is seen that this pattern receives very little attention in any of the four grades. Even when the "total repetitions" in all four grades combined are considered this pattern receives only 19 repetitions in Text X and a maximum of 29 repetitions in Text Z. This is a small offering spread over a period of four years. The questions in problems of Pattern D3 are always concerned with finding the length of time required to save a desired amount of money, to do a certain task, or to use supplies on hand, when the rate of saving, working, or using is given. It would seem that problems of this pattern are met frequently enough in life activities to warrant more emphasis than they have received in these texts.

Division Patterns D4, D5, and D6

The second important group of division patterns which represent the measurement concept of division is made up of those patterns which involve concepts of *distance, rate,* and *time.* There are three patterns in this group, namely, D4, D5, and D6. A glance at the graphic chart of Pattern D4, given on page 55, shows no treatment whatever of this pattern by any text in Grade 3 and practically no treatment in Grade 4 except by Text Z. It is easy to understand why this pattern receives so little attention in Grades 3 and 4 since Pattern D4 represents concepts that are rather difficult for pupils of these two grades. Fair attention is given this pattern in Grade 5, with wide variation in emphasis by the four authors. In Grade 6, however, where we might expect this pattern to have more liberal attention, we find that it has less attention than in Grade 5 on the part of three authors and no attention whatever on the part of Author W. Another outstanding fact evident from the graphs of D4 is that this pattern occurs almost exclusively in one-step problems.

The graphic charts of Patterns D5 and D6 are given together on page 56. It is very evident that both these patterns are almost non-existent in the four texts. It is easy to understand why these patterns should have little or no representation in Grades 3 and 4, and even in Grade 5, but it is not clear why they do not have more attention in Grade 6. One wonders whether these patterns slipped in without any plan on the part of the authors. D6 is a pattern of rare occurrence in the four texts examined, yet problems of this pattern can readily occur in life situations.

It will be remembered that all three patterns, D4, D5, and D6, involve concepts which are difficult for many pupils. Furthermore, expressions such as *at the rate, average speed, averages so many miles an hour, averages so many miles on one gallon,* etc., found in problems of these patterns, are very technical and add to their difficulty. The concepts involved in these three patterns are much more difficult than those involved in Patterns D1, D2, and D3, yet the latter together receive much greater emphasis than the former. The disparity in emphasis on these two groups of patterns raises several interesting questions. Would it be desirable to increase the offering of Patterns D4, D5, and D6, particularly in Grade 6? Are the concepts of these patterns of such difficulty that they might well be postponed until later grades? Should the few illustrations of these patterns now included in Grades 5 and 6 be considered as suitable only for the more able pupils, leaving the more detailed treatment for all pupils for later grades? Should each of the three patterns receive equal emphasis?

Division Patterns D7 and D8

The third group of patterns illustrating the measurement concept of division consists of two patterns, namely, Patterns D7 and D8. These two patterns involve the comparison of two numbers. In Pattern D7 the comparison is indicated by such questions as *how many times as large, how many times as many,* etc., whereas in Pattern D8 the comparison is indicated by the question *what part of.* The following problem illustrates Pattern D7: "Tom has saved $75 and Joe has saved $15. Tom's savings are *how many times as large* as Joe's?"

The graphic chart for Pattern D7 is given on page 57. The graph for Grade 3 shows no treatment whatever of this pattern by Texts X, Y, and W and only one exercise by Text Z. This is easily explained since division has relatively little treatment in Grade 3 by any of the texts. In Grades 4 and 5 there is a very wide variation in the offerings, indicated both by the graphs and by the unusually high coefficients of variation. The graphs for Grade 6 and for the four grades combined are slightly better, but even here the "one-step repetitions" vary greatly.

It is an accepted fact that comparison by means of division, like comparison by means of subtraction, is difficult and it is only natural to see relatively few comparison problems before Grade 5.

One reason for the variation in the findings on Pattern D7 is that the authors differ in their method of treating a two-step problem like the following: "When berries sell at 2 quarts for 35¢, how much must I pay for 6 quarts?" In solving this problem some of the authors follow the method of *unitary analysis* while others give the *ratio* method. By the method of unitary analysis the solution is as follows: "If 2 quarts cost 35¢, 1 quart will cost ½ of 35¢, or 17½¢. Then 6 quarts will cost 6 × 17½¢, or 105¢." It is seen that this particular solution makes no use of Pattern D7. By the ratio method the solution is like this: "Since 6 quarts are 3 times as much as 2 quarts, 6 quarts will cost 3 times as much as 2 quarts, or 3 × 35¢ which equals 105¢." In this solution Pattern D7 is used. In analyzing problems of the type just mentioned to determine the patterns that were used, it has been the writer's policy to select the patterns in terms of the method used by the authors in their model solutions. Hence problems of the type under discussion were credited to Pattern D7 when the authors used the ratio method, but were not so credited if the method of unitary analysis was used. The analysis of the texts showed that Authors X, Z, and W used the ratio method in their model solutions, whereas Author Y solved these problems by unitary analysis.

Pattern D8 is also concerned with the comparison of two numbers, but in this case one has to determine *what part* the smaller number is of the larger number. This pattern really represents the ratio concept of a fraction, which is by far the most difficult concept connected with the study of fractions. In general, problems representing Pattern D8 are more difficult than those representing Pattern D7.

The graphic chart for Pattern D8 appears on page 58. It is seen that this pattern is non-existent in Grade 3 and substantially so in Grade 4, which is to be expected. The ratio concept of a fraction, represented by Pattern D8, is not studied by most pupils until the latter half of Grade 5, hence it is unreasonable to expect much appearance of this pattern before

Grade 5. The graphs show that Pattern D8 has a fair representation in Grade 5 in Texts Y and W and very few appearances in Texts X and Z. In Grade 6 there is also considerable variation, Text Y having twice as many "total repetitions" as Text X. The graph for all four grades combined also shows much variation.

Since Pattern D8 is more difficult than Pattern D7 it will be of interest to compare the graphs of these two patterns with respect to the "total repetitions" in all four grades combined. This comparison shows that Texts Y and W each give more attention to D8 than to D7, whereas Texts X and Z each give more attention to D7 than to D8. That is, two of the authors reverse the policy of the other two. In view of the importance of Patterns D7 and D8, it does not seem likely that the four authors would let their offerings of these two patterns remain as they now are if they were aware of the facts disclosed by the graphs of these patterns.

Division Patterns D9, D10, D11, and D12

The fourth important group of division patterns, which is concerned with finding the amount, the size, or the cost of each part, involves the *partition concept* of division. This group is made up of four patterns, namely, D9, D10, D11, and D12. Pattern D9 is the one in which the fractional part of the number to be found is actually expressed in fractional form, as ½, ⅓, etc., or in words, as a *half of* or a *third of.* A glance at the graphic chart of this pattern on page 59 shows that Pattern D9 is found only in Grades 3 and 4. That is because this type of problem is considered as representing a division pattern only in Grades 3 and 4, whereas in Grades 5 and 6 it is treated as a multiplication pattern. It will be remembered that in Grades 5 and 6 the definition of multiplication is broadened to include finding a fractional part of a number, hence Pattern D9 in Grades 3 and 4 becomes Pattern M5 in Grades 5 and 6 in this study.

The graph of D9 for Grade 3 shows very close agreement on "total repetitions," which are practically all in one-step form, on the part of Texts X, Z, and W. Text Y, with 20 repetitions, has a little more than half as many as each of the other three texts. The graph for Grade 4 explains itself.

The other three patterns comprising this group of partition patterns are D10, D11, and D12. Pattern D10 is concerned with those problems where one is given a *total amount* and the *number of parts* into which it is to be *equally* divided and is required to find the *size* of each part. In this pattern the number of parts is expressed by a whole number rather than by a unit fraction, as is the case in Pattern D9. Pattern D10 applies to a wide variety of problems but excludes those problems that belong to Pattern D11 (finding each one's share) or to Pattern D12 (finding the cost of one).

The graphic chart of Pattern D10 is given on page 60. This chart shows wide variations in the treatment of this pattern in Grades 3 and 5, with more uniform conditions in Grades 4 and 6. Attention is called to the falling off in the treatment of this pattern in Grade 6. Another outstanding fact is the liberal number of repetitions of this pattern in one-step problems in each grade.

Pattern D11 is concerned with sharing. The graphic chart of Pattern D11, given on page 61, shows extreme variation among the four texts in "total repetitions" in each of the four grades. These grade-to-grade variations are only partially evened up in the combined graph of all four grades, where one finds "total repetitions" ranging from 33 to 132 appearances. One also finds that this pattern is represented very infrequently in multiple-step problems. In studying the combined graph one wonders whether two of the authors are aware of their relatively small offerings and if the other two are aware of their relatively large offerings. It would seem that this pattern would deserve more attention on the part of certain authors in Grades 5 and 6, where social situations involving this pattern may occur in considerable variety.

Pattern D12 requires finding the *cost of one article* when the *total number of articles* of like kind and *their total cost* are given. The graphic chart of this pattern appears on page 62.

This pattern has small representation in Grade 3, as might be expected. There is a more liberal representation of this pattern in the other three grades with considerable variation in each grade. In the combined graph for all four grades the discrepancies of the several grades are largely eliminated. It is difficult to understand why Pattern D12 has considerably more attention in Grades 5 and 6 than is given to Pattern D11 in these same two grades.

There are some interesting contrasts in the treatment of Patterns D10, D11, and D12. If one turns to the graphic charts of these three patterns, given on pages 60, 61, and 62, and studies the graphs of all four grades combined, one finds for Text X that the "total repetitions" for D10 and D12 together are about equal to the offering for D11 alone; at the opposite extreme Text W gives almost 4 times, and Text Y almost 6 times, as much attention to D10 and D12 together as to D11. If the total offering for D10, D11, and D12 is combined for each text, it is evident that there is less variation among the texts in the total offering of the three patterns than in the offering of the separate patterns. In their underlying thought, these patterns are alike, for each requires finding a fractional part of a number.

Division Pattern D13

Pattern D13, which represents the partition concept of division, is concerned with finding an average. The findings concerning Pattern D13 are given in graphic form on page 63. In Grade 3 it is seen that this pattern is non-existent, with one exception. The surprising thing is that any text should have an exercise of this kind in Grade 3. The notion of an average is not an easy one for children, and it is usually first presented in Grade 4. The graphs for Grades 4 to 6 explain themselves. The combined graph for all four grades shows considerable uniformity with respect to this pattern on the part of three authors. In each grade multiple-step problems are in evidence in each text, which is to be expected since average problems often involve two or more steps.

Division Patterns D14, D15, and D16

The last group of partition patterns is made up of three patterns, namely, D14, D15, and D16. These patterns are concerned with finding an average in relation to distance, rate, and time. In D14 the average speed per hour or the average rate traveled per hour is required. From the graph of Pattern D14, given on page 64, it will be noted that this pattern does not appear in Grade 3; this is to be expected since it is too difficult for this grade. This pattern does appear, however, with wide variations, in Grades 4, 5, and 6 and these appearances are mostly in one-step form. The combined graph for the four grades shows a wide variation among the four authors in the "total repetitions" of this pattern, these repetitions ranging from 27 in Text Z to 79 in Text X. The concept of an average, coupled with the concepts of distance, rate, and time, is difficult for children even in the upper grades and the high school. There are many opportunities in Grade 6 in connection with the division of decimals for average problems relating to travel by airplane and motor car. In Grade 6, however, three of the authors have taken little advantage of these opportunities.

The graphic charts of Patterns D15 and D16 are found together on page 65. The problems of D15 require the average distance traveled on one gallon of gasoline, while those of D16 require the average time to travel one mile. A glance at the graphs of these two patterns shows that they are almost nonexistent so far as the four texts under consideration are concerned. Pattern D15, with one exception, does not occur before Grade 5, where it appears a total of 1 to 7 times. It appears in three of the texts in Grade 6. In all four grades combined this pattern appears only 11 times in Text X, 5 times in Text Y, 8 times in Text Z, and twice in Text W. In other words, Pattern D15 occurs only 26 times in *all four texts combined.* This is indeed a meager showing for a pattern as difficult as this.

The showing of Pattern D16 is still worse, as a glance at the chart on page 65 shows. In all four grades combined it appears once in Series X, 3 times in Series Y, not at all in Series Z, and 9 times in Series W. In other words, Pattern D16 occurs a total of only 13 times in four years in three of the four texts examined. This is such a small number of appearances that one wonders whether the authors consciously planned its inclusion in their texts.

Patterns D14, D15, and D16 are the partition counterparts of Patterns D4, D5, and D6 of the measurement group of division patterns. A comparison of the graphs of D4, D5, and D6 on pages 55 and 56 with those of D14, D15, and D16 will show some interesting parallels. Pattern D14, which is the partition counterpart of D4, receives more than 4 times as much emphasis as Pattern D4 in Series X, twice as much in Series Y, about ¾ as much in Series Z, and more than 7 times as much in Series W. Pattern D15 receives a little more emphasis than its measurement counterpart, Pattern D5, in three of the series. Pattern D16, like its measurement counterpart, Pattern D6, is found in only three of the series and so very infrequently as to be negligible. In Series W, Pattern D16 appears 9 times while Pattern D6 is omitted entirely; in Series Z, Pattern D16 is omitted while Pattern D6 appears 3 times. In Series X and Y Patterns D6 and D16 appear at least once.

The reader will recall that all three patterns, D14, D15, and D16, not only involve concepts which are difficult but also contain such technical expressions as *average speed per hour, average rate, average number of miles per gallon,* etc. As a group they receive less emphasis than the other two groups of partition patterns. Since the concepts in these patterns are so difficult and since these patterns receive less emphasis than the easier patterns, one naturally raises several questions. Would it be desirable to increase the offerings of D14, D15, and D16, or should the introduction of these patterns be delayed until later grades? Are all of these patterns of equal importance, or should more emphasis be given to D14 than to D15 and D16, as has been done in these texts? Should enough emphasis be given these patterns to enable pupils to solve them with facility, or should a few illustrations of these patterns be included in Grades 5 and 6 for the more able pupils, leaving the more detailed treatment for Grades 7 and 8?

Comparisons and Relations Between Patterns

Attention is now directed to such relations as may exist between the different division patterns as disclosed by an analysis of the four series of arithmetics. In Table III on page 86, the division patterns have been classified into six major groups; Groups I to III represent the *measurement concept* of division while Groups IV to VI represent the *partition concept* of division. These six groups are as follows: (I) the simpler patterns requiring that a number of groups be found; (II) the patterns involving distance, rate, and time; (III) the patterns involving the comparison of two numbers; (IV) the patterns involving the amount, size, or cost of each part; (V) the pattern involving an average not related to distance, rate, and time; and (VI) the patterns involving an average in relation to distance, rate, and time.

Table III shows some interesting facts. In Group I it is seen that Text X has a total of 255 repetitions of the patterns making up this group; corresponding figures are 272 for Text Y, 448 for Text Z, and 289 for Text W. It is seen that the wide variations in "total repetitions," which were found when the individual patterns were discussed on the preceding pages, are also present when these patterns are combined in groups. In Group I the 448 repetitions of Text Z are 80% more than the 255 repetitions of Text X. In terms of percentages Text X devotes about 29% of its total offering in division to Group I; corresponding percentages for Group I for the other texts are 34% for Text Y, 41% for Text Z, and 31% for Text W. In other words, each of the texts gives about one third or more of its entire offering in division to Group I.

Concerning Groups I, II, and III, which represent the measurement concept, it is seen that Group I ranks first, Group III second, and Group II third. So far as Group II is concerned, each text gives this group less than one tenth the attention given Group I.

Let us now study Groups IV, V, and VI relating to the parti-

TABLE III

DIVISION PATTERNS ARRANGED IN SIX MAJOR GROUPS

Pattern	*Text X*	*Text Y*	*Text Z*	*Text W*
Measurement Concept				
Group I				
D1 + D2 + D3	255	272	448	289
Simpler Patterns	29.4%	33.8%	41.4%	31.4%
Group II				
D4 + D5 + D6	20	20	42	10
Distance, Rate, Time	2.3%	2.5%	3.9%	1.0%
Group III				
D7 + D8	94	98	114	121
Comparison	10.9%	12.2%	10.5%	13.2%
Partition Concept				
Group IV				
D9 + D10 + D11 + D12	301	276	379	358
Size of Each Part	34.7%	34.3%	35.0%	39.0%
Group V				
D13	106	98	65	86
Average	12.2%	12.2%	6.0%	9.4%
Group VI				
D14 + D15 + D16	91	40	35	55
Distance, Rate, Time	10.5%	5.0%	3.2%	6.0%
Total Offering	867	804	1083	919

The number 255 above shows for Text X the total repetitions in all four grades of the patterns in Group I. The 29.4% shows what per cent 255 is of 867 which represents the total offering in division in Text X.

tion concept of division. Group IV is the outstanding group of these three and represents one third or more of the entire offering in division of each text. Group V holds the second place in order of importance, while Group VI holds the third place.

If we examine all six groups in relation to each other, it is found that Groups I and IV each represent about one third, or slightly more, of the entire offering of each text. In other words, these two groups together cover about *two thirds* of the entire offering in division in each text. This is not surprising since these two groups each cover a wide variety of situations, whereas the other four groups are each more limited in their applications and somewhat more technical in their phraseology.

TABLE IV

SUMMARY OF THE TOTAL REPETITIONS OF EACH DIVISION PATTERN IN ALL FOUR GRADES COMBINED

Text	*Item*	*D1*	*D2*	*D3*	*D4*	*D5*	*D6*	*D7*	*D8*	*D9*	*D10*	*D11*	*D12*	*D13*	*D14*	*D15*	*D16*	*Total*
X	Number of repetitions	149	87	19	17	2	1	56	38	65	49	120	67	106	79	11	1	867
	Per cent	17.2	10.0	2.2	2.0	0.2	0.1	6.5	4.4	7.5	5.7	13.8	7.7	12.2	9.1	1.3	0.1	100
	Rank	1	4	11	12	14	15	8	10	7	9	2	6	3	5	13	15	
Y	Number of repetitions	163	88	21	16	3	1	24	74	47	82	33	114	98	32	5	3	804
	Per cent	20.3	10.9	2.6	2.0	0.4	0.1	3.0	9.2	5.8	10.2	4.1	14.2	12.2	4.0	0.6	0.4	100
	Rank	1	4	11	12	14	16	10	6	7	5	8	2	3	9	13	14	
Z	Number of repetitions	276	143	29	34	5	3	68	46	80	55	132	112	65	27	8	0	1083
	Per cent	25.5	13.2	2.7	3.1	0.5	0.3	6.3	4.2	7.4	5.1	12.2	10.3	6.0	2.5	0.7	0.0	100
	Rank	1	2	11	10	14	15	6	9	5	8	3	4	7	12	13	16	
W	Number of repetitions	174	92	23	6	4	0	54	67	75	114	58	111	86	44	2	9	919
	Per cent	18.9	10.0	2.5	0.6	0.4	0.0	5.9	7.3	8.2	12.4	6.3	12.1	9.4	4.8	0.2	1.0	100
	Rank	1	4	11	13	14	16	9	7	6	2	8	3	5	10	15	12	

Group II of the measurement concept is the counterpart of Group VI of the partition concept, both these groups dealing with the difficult concepts of *distance, rate,* and *time*. It is surprising, in a way, that Group VI is given considerably more emphasis than Group II in three out of the four texts. In terms of percentages, Text X gives 10.5% to Group VI against 2.3% to Group II; Text Y give 5.0% to Group VI against 2.5% to Group II; Text W gives 6.0% to Group VI against 1.0% to Group II. Only in Text Z do these two groups have approximately equal attention.

A comparison of the total number of repetitions of all the *partition* patterns with the total repetitions of all the *measurement* patterns shows that the partition notion of division receives a little more emphasis than the measurement notion in Series X, Y, and W, whereas in Series Z the measurement notion receives a little more emphasis than the partition notion.

We will discuss next such relations as may exist between the separate patterns. Table IV, on page 87, shows for each text the total number of repetitions of each pattern in all four grades combined. In all four texts, Pattern D1 receives the greatest emphasis. The emphasis given Pattern D1 ranges from 17.2% of the total offering in division in Text X to 25.5% in Text Z. There is little or no agreement as to the ranks of the other more important patterns. There is fairly close agreement, however, with respect to several patterns that rank very low (D5, D6, D15, and D16), the ranks of these patterns varying in each text from 12 to 16, 16 being the bottom of the list.

The total repetitions of each pattern, as given in Table IV, are shown in graphic form in Chart 2 on page 66. This chart makes it possible to compare the treatment of the sixteen division patterns in any one text with the corresponding treatment in each of the other texts.

For purposes of reference, Tables XVIa, XVIb, XVIc, and XVId will be found in the Appendix. These tables summarize the complete findings with respect to the sixteen division patterns for each of the four texts examined. The graphic charts given in this chapter were based upon data in these tables.

CHAPTER V

MULTIPLICATION PATTERNS

MULTIPLICATION OF INTEGERS is closely related to the process of addition and may be defined as a shortened process of finding the sum of a number of equal addends. This is only a partial definition of multiplication but it is a simple one and the one taught first. Multiplication takes on a broader meaning in the later grades when it is applied to the multiplication of common fractions and decimals.

An analysis of all the multiplication problems found in the four series of arithmetics reveals eight patterns. Five represent the simpler concept of multiplication mentioned above, while the other three are concerned with the more extended concept of multiplication. Some of the patterns are so related in their basic thought that they can be grouped together, thus making five major groups of patterns. These groups are as follows: (I) The patterns that require finding the total amount, exclusive of distance; (II) the patterns requiring the total distance and involving the concepts of rate and time; (III) the patterns requiring a fractional part of a number to be found; (IV) the pattern requiring a per cent of a number to be found; and (V) the pattern requiring a number to be found when a part of it is given. The discussion of these groups follows.

I. THE PATTERNS REQUIRING THE TOTAL AMOUNT TO BE FOUND

M1. Given the *number* or *amount in one group* and the *number of groups,* find the *total number* or *amount.*

Characteristic expressions, such as *in all, both, all together, twice as much, two times as many,* etc., that help emphasize the finding of a total, appear in some of the problems of this pattern.

(1) Bob has 2 bags of marbles. He has 45 marbles in each bag. How many marbles has Bob *in all*?

(2) Each page of Bill's stamp book holds 64 stamps. He has 16 pages full of stamps. How many stamps has he *all together*?

(3) Harry delivers 156 newspapers every week. How many papers will he deliver in 35 weeks?

(4) Pauline has a pudding recipe which uses ¾ cup of milk. She wants to make *3 times as much* pudding as the recipe provides. How much milk should she use?

M2. Given the *cost* or *amount of one unit* and the *number of units,* find the *total cost* or *amount of money.* Such words as *cost, charge, pay, spend, earn,* etc., are characteristic of this pattern. Emphatic words, such as *in all, both, all together, twice as many,* etc., appear in some of the problems of this pattern.

(1) Ned bought 4 tickets to the circus. Each ticket cost 75¢. How much did he *pay* for them?

(2) Bob earns 75¢ a week working for his father. How much will he *earn* in 2 weeks?

(3) If a baseball suit costs $4.85, how much in all will 9 baseball suits *cost*?

The underlying thought in each of these two patterns is that of finding the total number or amount when the number or amount in one group and the number of groups are given. Pattern M1 is very general and includes all those problems which do not specifically belong to Pattern M2. The latter includes only those problems which require finding the total cost or the total amount of money.

The second major group of patterns consists of those patterns which involve concepts of distance, rate, and time. They are generally considered more difficult than the first group.

II. THE PATTERNS REQUIRING THE TOTAL DISTANCE AND INVOLVING THE CONCEPTS OF RATE AND TIME

At the rate, average speed, averages so many miles an hour are technical expressions characteristic of this pattern.

M3. Given the *rate per unit* of *time* and the *time,* find the *total distance.*

(1) At 6 miles an hour, how far can you go on your bicycle in 3 hours?

(2) A very fast train ran for 6 hours at an average of 84 miles an hour. How far did it go in the 6 hours?

(3) How far will an airplane go in 3½ hours if its *average speed* is 150 miles an hour?

M4. Given the *mileage per gallon* of gasoline and the *number of gallons* of gasoline, find the *total distance.*
Miles per gallon, at the rate, averages so many miles on 1 gallon are technical expressions characteristic of this pattern.

(1) Our automobile can travel 16.5 miles on one gallon of gas. *At that rate,* how far can it travel on 2.5 gal.?

(2) Mr. Wells says that his car *averages* 18.4 mi. *per gallon of gasoline.* How far can he go on 15 gal.?

In these two patterns, M3 and M4, as in Patterns M1 and M2, the basic idea is finding the total amount, but the concepts involved in these patterns are more difficult than those in Patterns M1 and M2. Pattern M3 is concerned with finding the total distance traveled in a given time when the *rate* or *average speed per hour* is known, while Pattern M4 is concerned with finding the total distance that can be traveled on a given number of gallons of gasoline when the *average mileage per gallon* is known. Such expressions as *rate, average speed, mileage per gallon,* etc., involve concepts which are very difficult for some children, yet understanding of these technical expressions is necessary for the solution of problems of these patterns. These two patterns, M3 and M4, are the multiplication counterparts of the division patterns D4, D5, D14, and D15 and involve the same technical expressions. It will be recalled that these four division patterns are among the most difficult division patterns.

In each of the four multiplication patterns thus far discussed, the multiplier is *larger than 1.* It may be an integer, a mixed number, or a mixed decimal. In the third major group of multiplication patterns which follows, the multiplier is *less than 1.* It consists of those patterns in which the multiplier is either a common fraction or a pure decimal.

III. THE PATTERNS REQUIRING THAT A FRACTIONAL PART OF A NUMBER BE FOUND

M5. Given the number, find a fractional part of it (the numerator of the fraction being 1, as ½, ⅓, ⅙, etc.). This pattern includes pure decimal multipliers like .1 or .01. (See note below.)

(1) Mary and Tom bought a bag of popcorn together for 12¢. Each paid ½ of the cost. How much did each one pay?
(2) John had 14 marbles. He gave his brother ½ of them. How many marbles did he give his brother?

M6. Given the number, find a fractional part of it (the numerator of the fraction being greater than 1, as ⅔, ¾, ⅚, etc.). Pattern M6 includes multipliers in pure decimal form like .6 or .78.

(1) A grocer charges 27¢ a pound for cookies. How much does he charge for ¾ pound?
(2) A bushel of potatoes weighs 60 lb. If .78 of a potato is water, how many pounds of water are there in a bushel of potatoes?

Note. Pattern M5 is considered as a multiplication pattern only in Grades 5 and 6, and above. In Grades 3 and 4 this pattern is treated as a division pattern since the pupils in those grades are taught to find ½ or ⅓ of a number by dividing the number by 2 or 3 as the case may be. When the pupils reach Grade 5 they learn a new concept of multiplication for cases where the multiplier is a fraction; this new concept is much too difficult for Grades 3 and 4.

The problems of these two patterns require multiplication by a fraction or by a decimal, which is really finding a fractional part of a number. This involves a new meaning of multiplication which is very different from the one associated with multiplication when the child is first learning this operation. Pattern M6, in which the numerator of the fraction is larger than 1, appears as a multiplication pattern for the first time in Grade 5. A few problems of this type do occur in Grade 4, but in that grade they are treated as two-step problems by all four authors. In Grade 5, however, all four texts treat this type of problem as a one-step problem, which explains the appearance of M6 as a new pattern in Grade 5.

The next group includes only one pattern, this pattern requiring a per cent of a number to be found. There is so much that is new in percentage that problems of this kind have been put in a pattern by themselves even though finding a per cent of a number may be considered merely another form of expression for finding a fractional or decimal part of a number.

IV. The pattern requiring a per cent of a number

M7. Given the number, find a per cent of it.

(1) The school nurse says that 20% of the children in one class are underweight. Since there are 25 children in the class, there must be ..... children underweight.

(2) Mrs. Cox gets 5% off on a gas bill of $2.40 for paying it promptly. How much reduction does she get?

Problems of this pattern present several difficulties, namely, those caused by the technical phraseology in which the problems are expressed, those due to lack of familiarity with the social situations involved, and those due to writing the per cent as a decimal or as a common fraction. Such expressions as *per cent, 5% off, reduction of 25%, commission, profit, discount,* etc., are very technical and must be thoroughly understood before the problems can be intelligently solved. In addition, the child must know how to write a per cent as a decimal or as a common fraction and vice versa.

The fifth group also consists of only one pattern, as indicated below.

V. The pattern requiring the whole when a part is given

M8. Given *one* part of the number, find the *total* number.

(1) Nancy paid 10¢ for ¼ lb. of butter. At this rate, how much would 1 lb. of butter cost?

(2) Harry bought a half dozen nut bars for 24 cents. How much would he pay for a dozen of the same kind?

(3) Bert spent 75¢ on lunches last week. This was ½ of his allowance. What was his allowance?

The problems of Pattern M8 are simple since they require the whole when the given part is a unit fraction, such as ½, or ¼. Such problems are one-step problems under this pattern. Problems in which the given part is a fraction other than a unit fraction do not come under this pattern since they are treated in Grade 6 as two-step problems; it is not until Grade 7 that such problems are considered one-step problems.

It is of considerable interest to note that all the 4,359 one-step multiplication elements which were found in the four series of texts examined could be grouped into eight different patterns. Of these 4,359 multiplication elements, 2,599 were one-step problems and 1,760 were the multiplication step in multiple-step problems. Eight patterns is a very small number, considering the large number of multiplication situations examined and the great variety of social situations and language expressions used in these problems.

As each one-step problem belonging to a specific pattern contributes to the formulation of a generalization about that pattern, so each pattern contributes to the formulation of certain larger generalizations regarding the process of multiplication. The problems of Patterns M1, M2, M3, and M4 contribute to the notion that multiplication is the process by which the total amount is found when the amount in one group and the number of groups are given. In a sense, Pattern M8 also contributes to this concept of multiplication. Problems of Patterns M5, M6, and M7 contribute to a different meaning of multiplication, that of multiplication by a fraction or decimal which is defined as finding a fractional part of a number. Wide experience with many problems under each of these patterns helps in the understanding of these broad generalizations regarding multiplication.

Distribution of the Eight Multiplication Patterns in Four Series of Arithmetics

At this point we shall discuss the findings on multiplication as presented in the four series of textbooks which were examined and note how the four authors have treated each of the multiplication patterns and what emphasis they have given to each of the broad meanings of multiplication.

Multiplication Patterns M1 and M2

Patterns M1 and M2 are the easiest multiplication patterns, being concerned with finding the total amount when the number in one group and the number of groups are given. The graphic chart of Pattern M1, given on page 100, reveals some interesting facts. In Grade 3 one finds that Texts X, Y, and W exercise Pattern M1 exclusively in one-step problems, while Text Z has all but one repetition of this pattern in one-step form. There is considerable variation in the total number of appearances of this pattern in this grade, the totals varying from 41 to 76.

The graph for Grade 4 shows close agreement with respect to the "total repetitions" of Pattern M1 in Texts X, Y, and Z, these totals varying from 84 to 88. Text W, with a total of 127 repetitions, has about 1½ times as many repetitions as each of the other three texts. If the "total repetitions" for Grades 3 and 4 are combined for each text, it will be seen that there is much closer agreement among the four texts than when each grade is considered separately. One-step problems account for by far the greater number of appearances of Pattern M1 in this grade in all four texts. Here also, as in Grade 3, there is fairly close agreement among Texts X, Y, and Z on the repetitions of M1 in one-step problems.

The graph for Grade 5 presents a situation somewhat different from that in Grade 4. The number of repetitions of Pattern M1 in one-step problems in each text is about half as many as in Grade 4, while the number of repetitions in multiple-step problems is from 2 to 3 times as many, except in Text X. In other words, Pattern M1 appears less frequently in one-step problems and more frequently in multiple-step problems than is true in Grade 4.

The graph for Grade 6 explains itself. The graph of all four grades combined discloses the fact that the "total repetitions" of Pattern M1 range from 247 to 350, the average for all four texts being 300 repetitions. None of the patterns previously studied in subtraction or in division has had such a high representation. In subtraction the pattern most frequently found was Pattern S1, which had an average of 197 repetitions in the four grades combined. In division Pattern D1 had the largest count, with an average of 190 repetitions in all four grades combined. It is apparent that the repetitions of Pattern M1 greatly exceed anything noted before. Pattern M2, which will next be described, greatly outstrips Pattern M1. The average for Pattern M2 in all four grades combined is 585 representations, being almost twice that of Pattern M1.

The reader will recall that Pattern M2 is concerned with finding the total cost or total amount of money. This pattern is closely related to Pattern M1, for in both patterns a total amount is to be found. In M1 the total amount may refer to anything except money, while in M2 the total amount refers to money only. M2 is kept as a separate pattern because of the technical vocabulary involved in the problems of this pattern and also the important social situation represented by it, namely, that of buying and selling.

The graphic chart of Pattern M2 is given on page 101. In Grade 3 all authors give a generous amount of repetition to this pattern, the total number of repetitions varying from 73 to 94. One-step problems account for all its appearances in Text W and by far the greater number of its appearances in Texts X, Y, and Z.

One striking feature of the graph for Grade 4 is the big increase in the number of repetitions of Pattern M2 in all four texts over Grade 3. Text X has about 1⅔ times as many as

Grade 3, Text Y about twice as many, Text Z about $2\frac{1}{3}$ times as many, while Text W has about $2\frac{1}{2}$ times as many. Another striking feature in this grade is the very liberal appearance of M2 in multiple-step problems, especially in Texts X and Z.

The graph for Grade 5 shows a decrease in the number of repetitions of Pattern M2 in one-step problems and an increase in its appearance in multiple-step problems, this increase being especially large in Texts Y and W. In this grade Pattern M2 appears most frequently in multiple-step problems. The graph for Grade 6 presents nothing especially new.

The graph for all four grades combined discloses the astonishing fact that all four authors have given this pattern a total number of repetitions ranging from 503 to 628, thus making it the pattern most frequently found in this entire study. The very liberal offering of this pattern on the part of each author naturally raises the question of whether it has been overemphasized at the expense of other patterns. One reason probably for the large offering of this pattern is its wide social use in everyday life.

Multiplication patterns M3 and M4

The second group of multiplication patterns is made up of two patterns, namely, M3 and M4. These are the patterns which require finding the total distance and involve the concepts of rate and time. In Pattern M3 the total distance traveled is required when the rate per unit of time and the total time are given; in Pattern M4 the total distance traveled is required when the mileage per gallon of gasoline and the number of gallons are given. The graphic charts of these two patterns are given together on page 102.

Referring to the graph for Pattern M3, it is evident that this pattern receives very little attention in any of the four grades. In Grade 3 this pattern appears in but two texts, X and Y, and then so infrequently as to be negligible. In Grade 4 the "total repetitions" vary from 3 to 12; in Grade 5, from 1 to 10; and in Grade 6, from 4 to 13. The combined graph for all four grades shows only 13 repetitions in Text Y, while the maximum of

34 is found in Text X. In each case, this is a very small offering spread over a period of four years. The graph for Pattern M4 shows that this pattern is inconsequential in all four series, appearing a total of only 9 times in all four years in all four series combined.

Earlier in this chapter attention was called to the difficulty of these two patterns because of the concepts involved and the technical phraseology used in the problems. Examination of all the problems of these two patterns shows that such expressions as *average speed, average miles per hour, mileage per gallon* of gasoline do not appear before Grades 5 and 6. The problems of Pattern M3 found in Grades 3 and 4 are easier than those found in Grades 5 and 6, but they are insufficient in number to give facility in solving problems of this pattern. For many fourth grade children, finding how far an automobile can travel in 3 hours if it travels 35 miles an hour presents real difficulties, for the child has difficulty in visualizing the conditions of the problems. Likewise, finding how far one can travel on a given number of gallons of gasoline, as in M4, is of little or no concern to the child for it has reality for him only in the discussions which he may hear among adults. Perhaps the difficulty of these two patterns partially explains their meager treatment on the part of all four authors. At any rate the slight attention given these patterns raises several questions: Are the concepts of Pattern M3 of such difficulty that it might well be postponed until later grades? Should perhaps a few illustrations of this pattern be included in Grades 5 and 6 for the more able pupils, leaving the more detailed treatment for later grades? Were the very few illustrations of M4 on the part of all authors intentional or accidental?

Multiplication Patterns M5 and M6

The third important group of multiplication patterns involves a new concept of multiplication which is applied in problems where one finds a fractional part of a number. This new concept is usually introduced in Grade 5, whereas the simpler concept of multiplication is taught in Grades 3 and 4.

MULTIPLICATION PATTERNS

MULTIPLICATION PATTERN M1

Given the Amount in One Group and the Number of Groups, to Find the Total Number or Amount

▭ Repetitions in one-step problems
▬ Repetitions in multiple-step problems

Grade 3

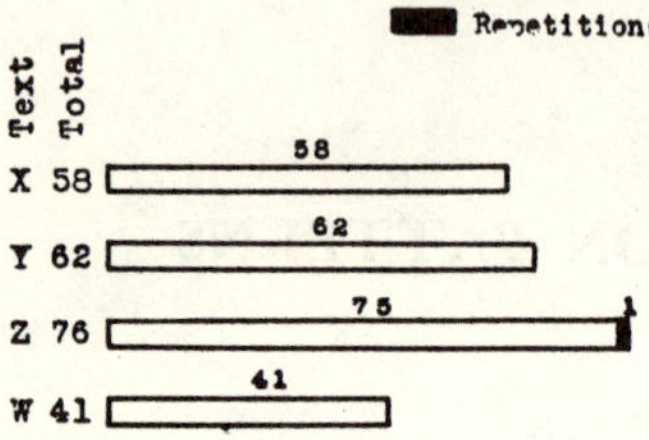

	1-Step	Total
σ	12.14	12.48
M	59.00	59.25
V	20.6	21.1

Grade 4

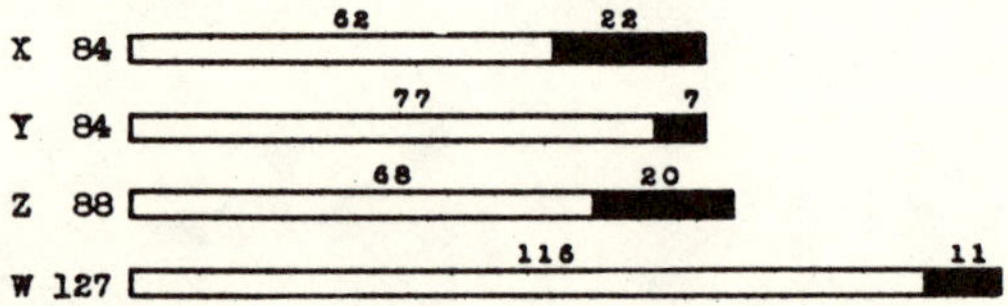

	1-Step	Total
σ	21.04	18.12
M	80.75	95.75
V	26.1	18.9

Grade 5

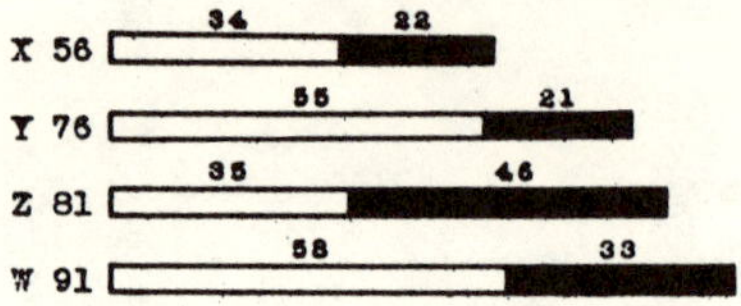

	1-Step	Total
σ	11.06	12.75
M	45.50	76.00
V	24.3	16.8

Grade 6

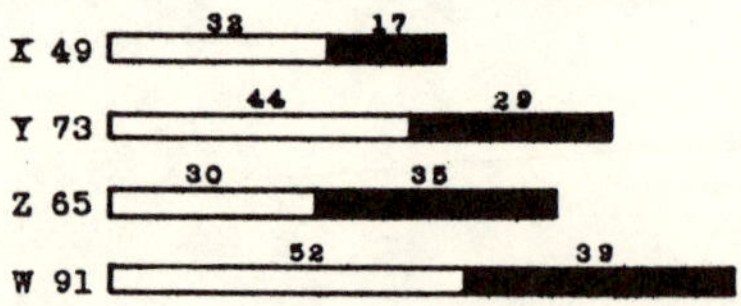

	1-Step	Total
σ	8.99	15.12
M	39.50	69.50
V	22.8	21.8

Grades 3 to 6 Combined*

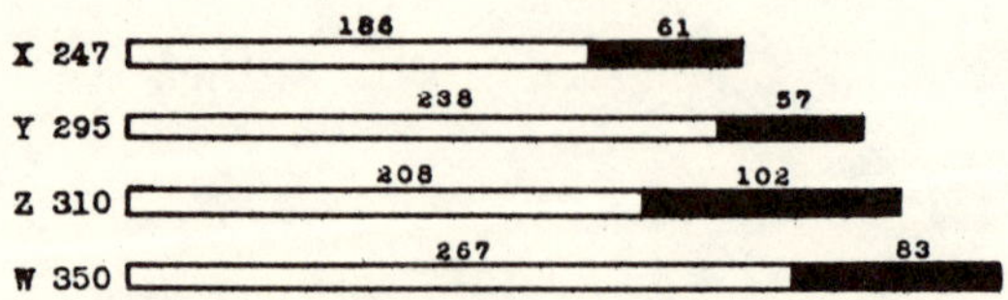

	1-Step	Total
σ	30.59	36.85
M	224.75	300.50
V	13.6	12.3

*The graph for Grades 3 to 6 Combined has been drawn to a reduced scale.

MULTIPLICATION PATTERN M2

Given the Cost of One Unit and the Number of Units, to Find the Total Cost

Grade 3

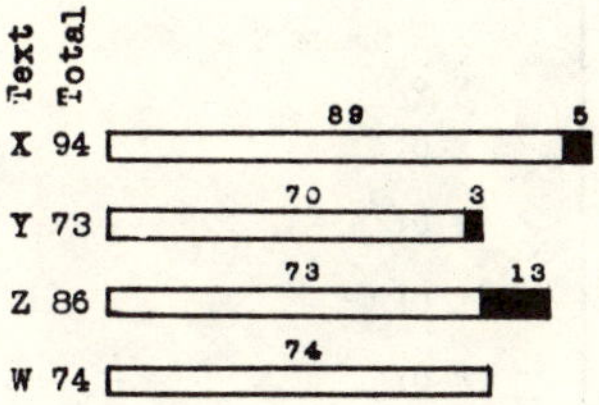

	1-Step	Total
σ	7.37	8.73
M	76.50	81.75
V	9.6	10.7

Grade 4

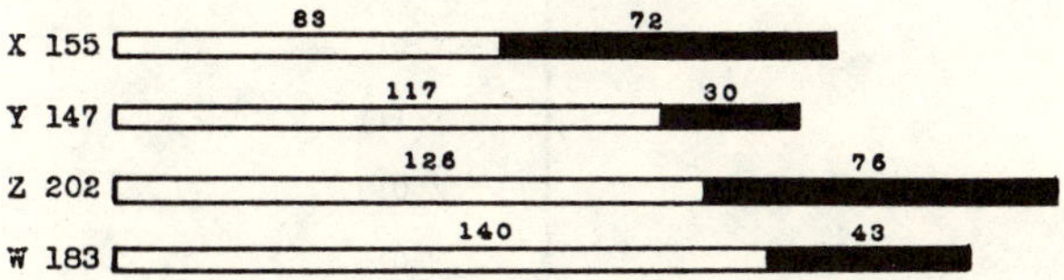

	1-Step	Total
σ	21.01	22.99
M	116.50	171.75
V	18.0	12.8

Grade 5

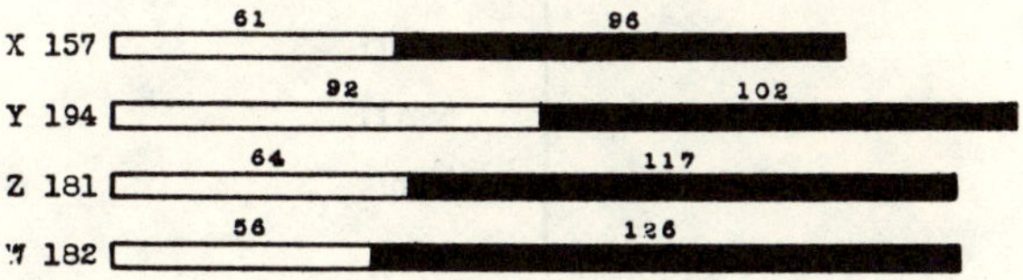

	1-Step	Total
σ	14.01	13.43
M	68.25	176.50
V	20.5	7.5

Grade 6

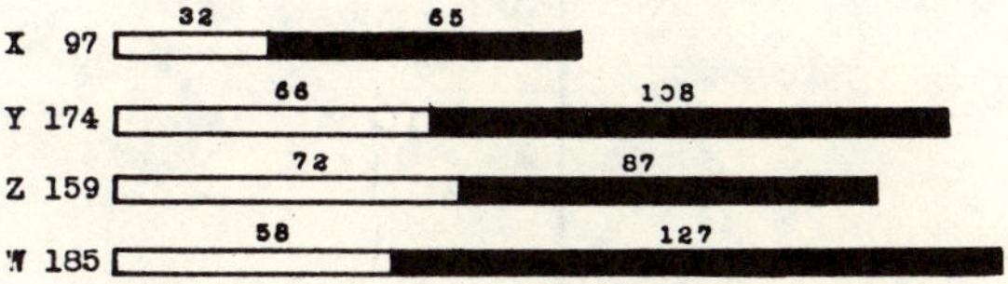

	1-Step	Total
σ	15.26	34.04
M	57.00	153.75
V	26.8	22.1

Grades 3 to 6 Combined*

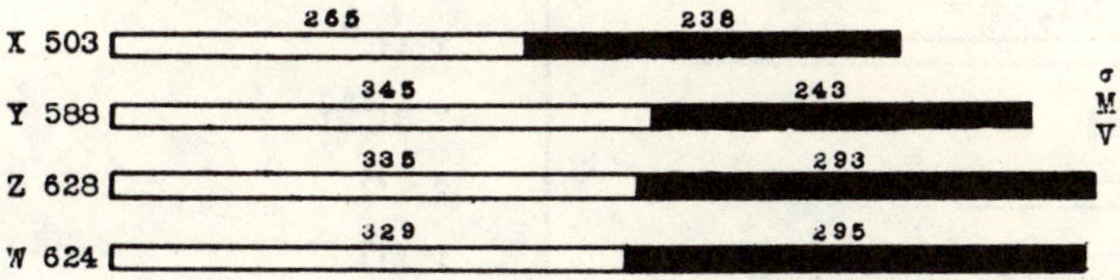

	1-Step	Total
σ	31.41	50.25
M	318.50	585.75
V	9.9	8.6

*The graph for Grades 3 to 6 Combined has been drawn to a reduced scale.

MULTIPLICATION PATTERN M3

Given the Rate and the Time, to Find the Total Distance

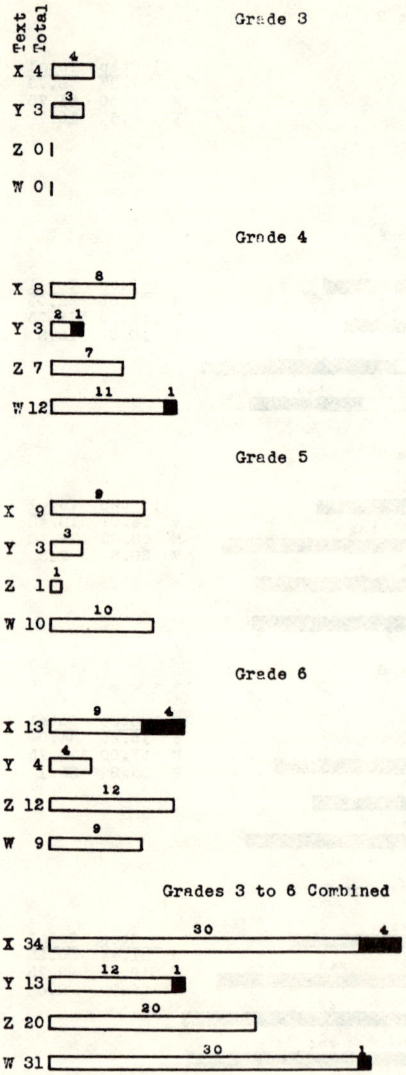

MULTIPLICATION PATTERN M4

Given the Mileage per Gallon and the Number of Gallons, to Find the Total Distance

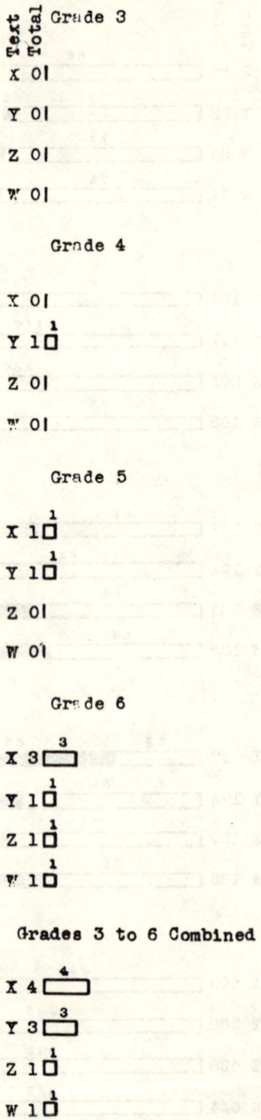

MULTIPLICATION PATTERN M5

Finding a Fractional Part of a Number, the Numerator of the Fraction Being One

Grade 3

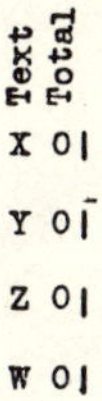

Grade 4

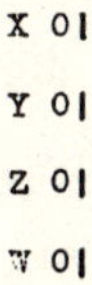

Grade 5

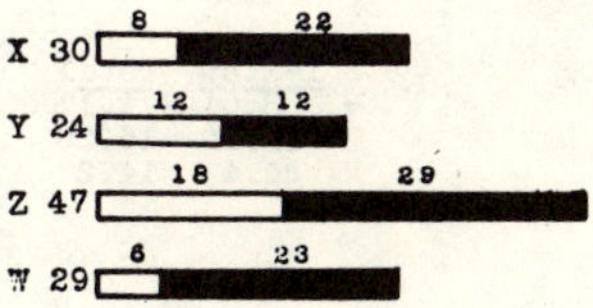

	1-Step	Total
σ	4.58	8.67
M	11.00	32.50
V	41.6	26.7

Grade 6

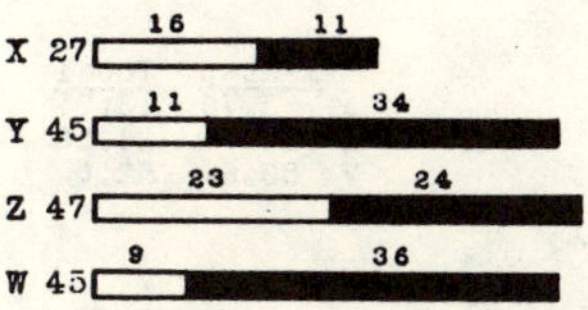

	1-Step	Total
σ	5.40	8.12
M	14.75	41.00
V	36.6	19.8

Grades 3 to 6 Combined

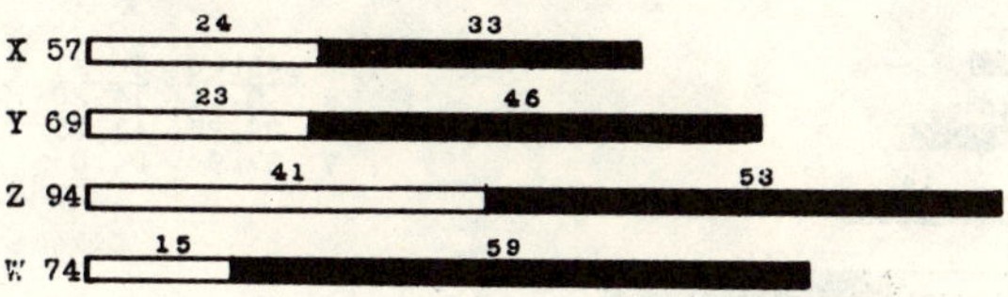

	1-Step	Total
σ	9.47	13.35
M	25.75	73.50
V	36.8	18.2

MULTIPLICATION PATTERN M6

Finding a Fractional Part of a Number,
the Numerator of the Fraction Being Greater Than One

Grade 3

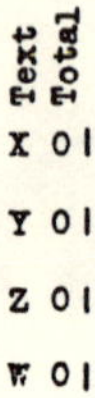

Grade 4

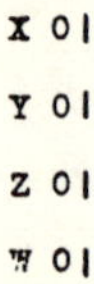

Grade 5

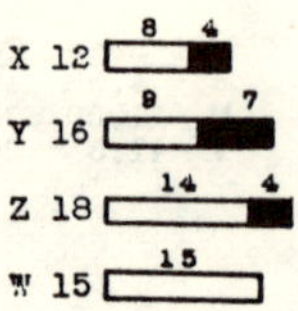

	1-Step	Total
σ	3.04	2.17
M	11.50	15.25
V	26.4	14.2

Grade 6

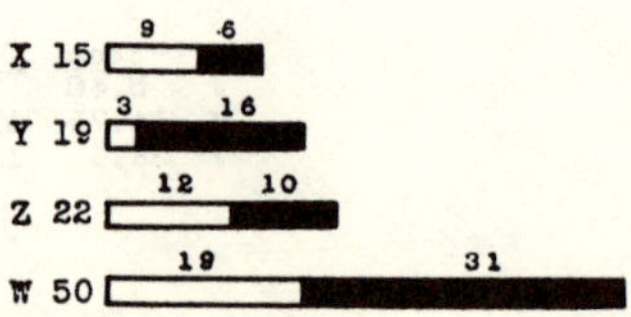

	1-Step	Total
σ	5.76	13.79
M	10.75	26.50
V	53.6	52.0

Grades 3 to 6 Combined

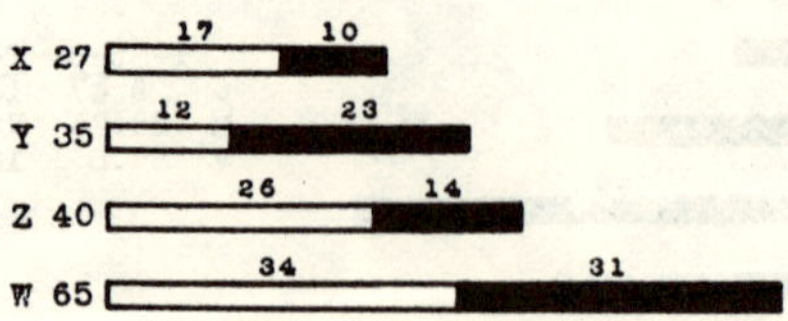

	1-Step	Total
σ	8.44	14.20
M	22.25	41.75
V	37.9	34.0

MULTIPLICATION PATTERN M7

Finding a Per Cent of a Number

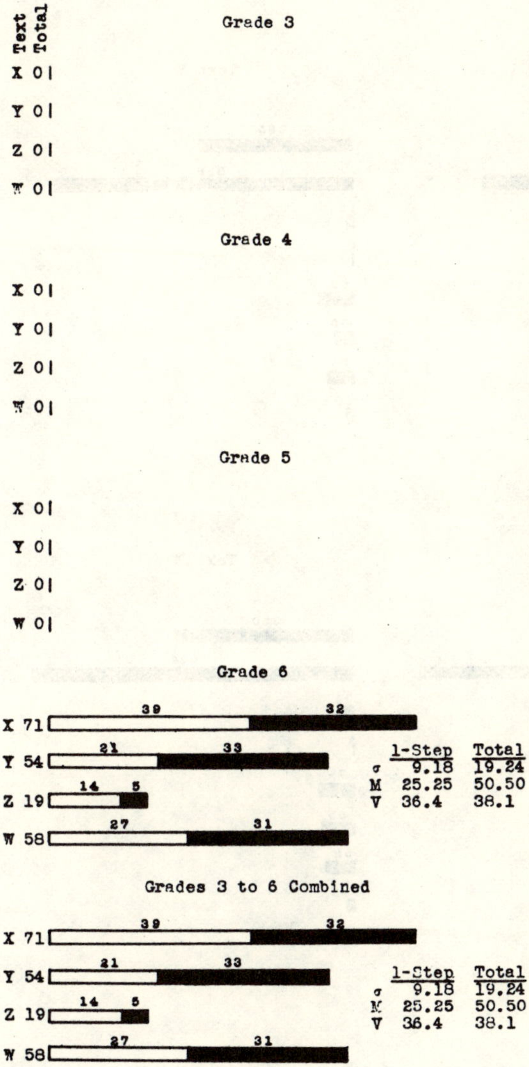

MULTIPLICATION PATTERN M8

Given a Part of a Number, to Find the Total Number

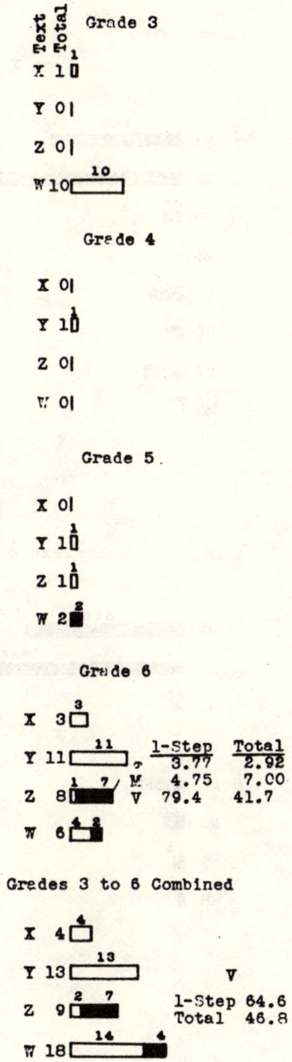

CHART 3

Total Repetitions of Each of the Multiplication Patterns in All Four Grades Combined, Arranged by Texts

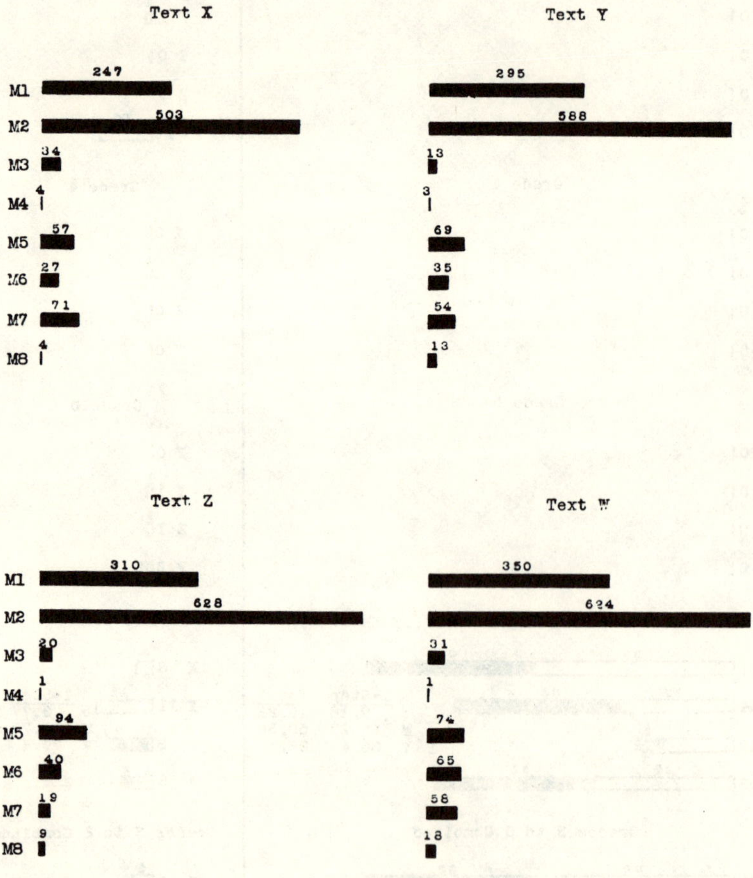

The simpler concept of multiplication is illustrated by the following problem: "How much do 2 apples cost at 5¢ each?" To solve this problem one has to find the value of $2 \times 5¢$. In this case the multiplier 2 is a *whole number* and it means that 5¢ is taken *2 times.* In other words, when the multiplier is a whole number it tells *how many times* the multiplicand is to be taken. The new concept of multiplication is illustrated by this problem: "Find the cost of 3/4 lb. of tea at 60¢ a pound." When this problem is presented in Grade 5 it requires one to find the value of $3/4 \times 60¢$; this is considered a one-step multiplication problem and 3/4 is called the multiplier. But this multiplier no longer tells *how many times* 60¢ is taken. To do anything "3/4 times" is meaningless. Hence, when the multiplier is a fraction, the definition of multiplication has to be enlarged so that $3/4 \times 60¢$ has meaning. The pupils are taught in a case like this that $3/4 \times 60¢$ means 3/4 of 60¢ and that the sign $\times$ in this situation means "of." This is the new concept of multiplication introduced in Grade 5. In this study a problem like $3/4 \times 60¢$ is considered a one-step problem in Grades 5 and 6 and is classified under Pattern M6.

Similarly, in Grades 5 and 6 a problem involving the evaluation of $1/2 \times 70¢$ is considered a multiplication problem, the fraction 1/2 being the multiplier. According to the enlarged notion of multiplication, $1/2 \times 70$ means 1/2 of 70. In this study, when the multiplier is a unit fraction like 1/2 or 1/3, the problem is classified under Pattern M5 in Grades 5 and 6. A problem in which one has to find 1/2 of 70¢ is also met in Grades 3 and 4 but in such cases it is always written as 1/2 of 70¢ and never as $1/2 \times 70¢$. In Grades 3 and 4 a problem involving 1/2 of 70¢ comes under Division Pattern D9, since one finds the result by dividing 70¢ by 2.

From the discussion above it is seen that Patterns M5 and M6 are each concerned with finding a fractional part of a number; they differ only with respect to the numerator of the fraction. In Pattern M5 the fraction has a numerator of 1; in Pattern M6 the fraction has a numerator greater than 1.

The graphic chart of Pattern M5 is shown on page 103. The

chart shows that this pattern does not appear at all in Grades 3 and 4; the reason for this has just been discussed in the preceding paragraphs, where it was pointed out that in Grades 3 and 4 a problem requiring one to find a unit fractional part of a number is considered a division problem and is classified under Division Pattern D9. The graphs of Pattern M5 for Grades 5 and 6, as well as the combined graph for these two grades, show that there is a wide variation in each grade in the number of one-step problems and that there is also a liberal appearance of this pattern in multiple-step problems. In Grade 5 three authors agree rather closely with respect to the "total repetitions" of this pattern; this same situation is also found in Grade 6.

The graphic chart of Pattern M6 appears on page 104. From this chart it is seen that this pattern is not represented in Grades 3 and 4 for reasons that have already been stated. The graph for Grade 5 shows close agreement with respect to "total repetitions" for all four texts but considerable disagreement with respect to the offering in one-step problems and multiple-step problems. In Grade 6 the discrepancies in the repetitions in one-step problems and in multiple-step problems are greater than in Grade 5. In Grade 6 there is also wide variation in "total repetitions"; it is seen that Text W has more than twice the offerings of Texts Y and Z and more than 3 times that of Text X. In the graph for both grades combined the discrepancies in the separate grades are evened up somewhat, but the same general tendencies are shown, namely, wide variation in "one-step," "multiple-step," and "total repetitions."

A comparison of the treatment of Pattern M6 with that of Pattern M5 will now be made. On the chart of Pattern M6, let us study the graph at the bottom of the chart which gives the total offering for Grades 5 and 6 combined; this graph shows that the "total repetitions" of this pattern in the four texts range from 27 to 65 with an average of 42 repetitions; in the corresponding graph for Pattern M5, the "total repetitions" range from 57 to 94 with an average of 74 repetitions. It is seen that Pattern M5 has about 75% more repetitions than M6.

One might expect this situation to have been reversed since Pattern M6 is less restricted than M5; Pattern M6 deals with any common fraction with a numerator greater than 1, whereas M5 is confined to unit fractions. It is granted, however, that unit fractions have a wide range of usefulness in practical life.

Multiplication Patterns M7 and M8

Pattern M7, which is the next pattern to be considered, is concerned with finding a per cent of a number. In the early part of this chapter, where this pattern was described, some of its difficulties were pointed out. Ability to solve problems of this pattern depends largely upon the understanding of the social situations involved and the technical phraseology used. Furthermore, computation in these problems depends upon a knowledge of decimals. Obviously this pattern is too difficult for early introduction in the elementary school and the writer seriously questions its introduction in Grade 6. A glance at the graphic chart of Pattern M7, which is found on page 105, shows that this pattern does not appear until Grade 6, hence the graph for all four grades combined is the same as the graph for Grade 6. One striking feature of this graph is the wide variation in "total repetitions" of this pattern among the four authors. This wide variation may be due to the difficulty of the pattern and to the fact that it is being introduced for the first time in Grade 6. Since some schools teach no percentage whatever in Grade 6, there was probably doubt on the part of some of the authors as to whether they would do anything at all with percentage in Grade 6.

There is one more pattern to be considered, namely, Pattern M8, which requires finding the whole when a part is given. Certain aspects of this pattern have already been discussed on page 94. The problems of this pattern are simple since they require the whole when the given part is a unit fraction. The graphic chart of Pattern M8, given on page 105, shows that Pattern M8 is of very infrequent occurrence in each grade in all four texts. Even considering the total number of appearances in all four grades combined, this pattern receives a mimimum of only 4 repetitions in Text X and a maximum of 18 repetitions in Text W. One wonders why one author gives 10 repetitions of this pattern in Grade 3 and only 8 repetitions in Grades 5 and 6 combined.

Comparisons and Relations Between Patterns

Table V gives data concerning the multiplication patterns when classified into five major groups as follows: (I) the patterns requiring the total amount to be found; (II) the patterns requiring the total distance to be found; (III) the patterns requiring a fractional part of a number to be found; (IV) the pattern requiring a per cent of a number to be found; and (V) the pattern requiring the whole to be found when one part of it is given.

TABLE V

MULTIPLICATION PATTERNS ARRANGED IN FIVE MAJOR GROUPS

Pattern	*Text X*	*Text Y*	*Text Z*	*Text W*
Group I M1 + M2 Finding Total Amount	750 79.3%	883 82.6%	938 83.7%	974 79.8%
Group II M3 + M4 Finding Total Distance	38 4.0%	16 1.5%	21 1.9%	32 2.6%
Group III M5 + M6 Finding a Fractional Part of a Number	84 8.8%	104 9.7%	134 12.9%	139 11.4%
Group IV M7 Finding a Per Cent of a Number	71 7.5%	54 5.0%	19 1.7%	58 4.7%
Group V M8 Finding the Whole If a Part Is Given	4 0.4%	13 1.2%	9 0.8%	18 1.5%
Total Offering	947	1070	1121	1221

The number 750 above shows for Text X the total repetitions in all four grades of the patterns in Group I. The 79.3% shows what per cent 750 is of 947 which represents the total offering in multiplication in Text X.

Referring to Table V, Group I is represented in the four texts by "total repetitions" ranging from 750 to 974. In terms of percentages Text X devotes about 79.3% of its total offering to the two patterns in Group I; corresponding percentages for the other texts are 82.6% for Text Y, 83.7% for Text Z, and 79.8% for Text W. This means that about *four fifths* of the entire offering in multiplication in each of the four texts is devoted to these two patterns.

Group II, which consists of Patterns M3 and M4, relates to finding the total distance. The treatment of this group is in striking contrast to that of Group I. The "total repetitions" of this group range from only 16 to 38. The per cents of the total offering are also very low, ranging from 1.5% to 4.0%. It is seen that all four authors consider Group II of little importance.

Group III includes Patterns M5 and M6, which are concerned with finding a fractional part of a number. In the four texts this group represents from 8.8% to 12.9% of the total offering in multiplication. Of the five groups of multiplication patterns, Group III is second in importance among the four authors, Group I holding the first place.

Group IV involves finding a per cent of a number; the attention it receives ranges from 1.7% to 7.5% of the total offering in multiplication, thus indicating a wide variation in the offerings of the four authors.

Group V requires the finding of the whole when a part of it is given. All four authors consider this group of least importance since it represents such a small part of the total offering in each text, the per cents ranging from 0.4% to 1.5%.

Attention is called to the fact that all four authors agree on the relative importance of each of the major groups of patterns. Group I ranks first; Group III, second; Group IV, third; Group II, fourth; and Group V, fifth.

The last line of Table V shows the total offering in multiplication in all four texts, this offering ranging from 947 to 1,221 repetitions. By contrast, all the division patterns range in "total repetitions" from 804 to 1,083 (see page 86). Since multiplication and division are inverse operations, it is interesting to note that multiplication has received more attention relatively than division in the total offerings of the four texts.

Let us now examine Table VI, below, which gives data concerning the separate patterns in multiplication when they are not arranged in groups. Table VI shows that in all four

TABLE VI

SUMMARY OF THE TOTAL REPETITIONS OF EACH MULTIPLICATION PATTERN IN ALL FOUR GRADES COMBINED

Text	*Item*	*M1*	*M2*	*M3*	*M4*	*M5*	*M6*	*M7*	*M8*	*Total*
X	Number of repetitions	247	503	34	4	57	27	71	4	947
	Per cent	26.2	53.1	3.6	0.4	6.0	2.8	7.5	0.4	100
	Rank	2	1	5	7	4	6	3	7	
Y	Number of repetitions	295	588	13	3	69	35	54	13	1070
	Per cent	27.6	55.0	1.2	0.3	6.4	3.3	5.0	1.2	100
	Rank	2	1	6	8	3	5	4	6	
Z	Number of repetitions	310	628	20	1	94	40	19	9	1121
	Per cent	27.7	56.0	1.8	0.1	8.4	3.5	1.7	0.8	100
	Rank	2	1	5	8	3	4	6	7	
W	Number of repetitions	350	624	31	1	74	65	58	18	1221
	Per cent	28.7	51.1	2.5	0.1	6.1	5.3	4.7	1.5	100
	Rank	2	1	6	8	3	4	5	7	

texts Pattern M2 has by far the greatest emphasis; this pattern represents more than 50% of the total offering in multiplication in all four texts. Comparing Pattern M2 with M1, M2 receives a little more than twice as much emphasis as M1 in Texts X, Y, and Z and a little less than twice as much in Text W. Pattern M5 holds third place in three texts. There is little agreement concerning the ranks of the other patterns until we reach the lowest ranks, where it is found that Pattern M8 holds seventh place in three texts while Pattern M4 holds eighth place in three texts.

When the patterns are arranged in groups, as shown in Table

V, there is complete agreement among the four authors as to the ranks of the five groups. In Table VI, however, there is no such close agreement when the patterns are considered individually except with respect to Patterns M2 and M1, which hold the first and the second places respectively.

The total repetitions of each pattern, as given in Table VI, are shown in graphic form in Chart 3 on page 106. This chart makes it possible to compare the treatment of the eight multiplication patterns in any one text with the corresponding treatment in each of the other texts.

For purposes of reference, Tables XVa, XVb, XVc, and XVd are given in the Appendix. These tables incorporate complete results of the analysis of the four texts with respect to the eight multiplication patterns.

CHAPTER VI

ADDITION PATTERNS

ADDITION IS USUALLY CONSIDERED the easiest of the four fundamental operations. The basic idea in all addition problems is that of combining two or more numbers of the same kind to find their sum. This statement might lead to the conclusion that there is but one addition pattern. A critical examination of the addition problems, however, reveals certain minor differences among them. These differences form the basis for classifying the problems under four patterns. The distinctions between these patterns are less clearly marked, for the most part, than the distinctions between the subtraction patterns or the division patterns. A description of the four addition patterns is given below. These are labeled Patterns A1, A2, A3, and A4.

All patterns require a sum or total amount to be found.

A1. The phraseology in which the problems are expressed helps to emphasize the idea of finding a sum.

Characteristic expressions such as *in all, both, total, together, all together, the two, both together,* are used in the problems. The word *more* also gives emphasis in certain problems where a given amount has been increased by another amount.

(1) Today Harry saw 5 robins, 2 blue jays, and 8 sparrows. How many birds did he see *all together*?
(2) Fred sold 86 papers today and John sold 74 papers. How many papers did *both* boys sell?
(3) George had 485 stamps. He bought 239 *more* stamps. How many had he then?

In the above problems the expressions such as *all together, both,* and *more* help to emphasize the *underlying thought* of the problem, which is that of *finding a sum.* These words are not to be used as "cues" which indicate the process to be used in solving these problems; they merely serve to emphasize the basic thought of these problems, which is to put numbers together to form a whole.

A2. The phraseology in which the problems are expressed does not contain characteristic words or expressions such as those listed in Pattern A1.

(1) If Jane takes 4 dolls to the party, Ann takes 5, Shirley takes 6, and I take 3, how many dolls will we take?
(2) I had $2.36 and earned $.47. Now I have ......
(3) Mary made 24 white paper flowers and 24 blue paper flowers. How many flowers did she make?

Careful reading of these problems shows that they are very similar to the ones listed under Pattern A1. In each of these problems numbers of the same kind are to be combined to find a whole, but there are no words such as *all together, both, in all, more,* and *total* to emphasize the basic thought. The only difference, therefore, between the problems of these two patterns is the omission, in Pattern A2, and the inclusion, in Pattern A1, of the words just mentioned.

A3. The phraseology of the problems is peculiar to the activity of buying and selling.

Characteristic expressions such as *charge, pay, spend, cost, total cost, total bill, total expenses,* are used in the problems.

(1) Jane ate her lunch at school today. She got soup for 7¢, bread and butter for 4¢, milk for 5¢, and ice cream for 8¢. How much did her lunch *cost?*
(2) Jane spent 25¢ for a doll and 22¢ for a teddy bear. How much did she *spend* for both?
(3) Ned paid 18¢ for his lunch and 25¢ for his sister's lunch. How much did Ned *pay* in all?

The problems of Pattern A3 are like those of Pattern A1 and Pattern A2 in that a sum is required. Some of the problems of this pattern also contain such characteristic expressions as *both, all together, in all,* etc., which are found in Pattern A1 while other problems are like those of Pattern A2 in that such words as *both, all together,* etc., are omitted. The problems of Pattern A3, however, are concerned with buying and selling, which involves a technical vocabulary peculiar to this activity. Such words as *spend, cost, total cost, expenses,* and *total bill* are frequently used and are characteristic of problems which describe buying and selling transactions.

A4. The phraseology of the problems is similar to that used in certain subtraction patterns.

(1) Harry had some pennies. He spent 5 pennies and had 12 pennies *left.* How many did he have at first?

(2) Fred has 263 stamps in his collection. His brother Joe has 45 stamps *more than* Fred. How many stamps has Joe?

(3) We have a new playground at school. Our old playground was 185 feet long, but this one is 65 feet *longer.* How long is the new playground?

(4) John weighs 58¼ pounds, which is 3⅜ pounds below his normal weight. What is his normal weight?

The problems of this pattern also require the finding of a sum, but they contain phrases often found in certain subtraction patterns. For example, in the above illustrations, the wording of the first problem is somewhat similar to that of Pattern S1, which requires finding the amount left after spending a certain sum. The wording of the second problem is somewhat similar to that of Pattern S5, which compares two numbers by finding how much *more* one is *than* the other; the third problem is similar to Pattern S6, which compares two numbers by finding how much *larger* one is *than* the other; while the last is similar to Pattern S8, which compares two numbers by finding their difference, although the word "difference" is not used in the problem. This similarity in phraseology between the problems of Pattern A4 and the problems of certain subtraction patterns might lead some pupils to subtract rather than to add if they had been taught to look for cues rather than the underlying thought in the problem.

It is significant that all of the 2,468 one-step addition elements, the total number found in the four series of texts analyzed in this study, could be grouped under four patterns. This is a very small number of patterns considering the wide variety of social situations and language expressions used in the problems. Of this total, 1,578 are one-step problems and 890 are the addition step in multiple-step problems.

Distribution of the Four Addition Patterns by Grades in Four Series of Arithmetics

A careful study of the findings in addition as presented in the four series of textbooks which were analyzed will show how the four authors have treated each of the addition patterns. Only four patterns were found for addition in contrast with ten patterns in subtraction, which is the inverse of addition.

Addition Patterns A1 and A2

The graphic chart of Pattern A1, given on page 125, shows that this pattern occurs most frequently in Grade 3 in all four series. It receives gradually diminishing emphasis, in general, in each succeeding grade. Pattern A1 appears only once in a multiple-step problem in this grade.

The graph for Grade 4 presents much the same general picture so far as "total repetitions" are concerned. In all four texts multiple-step problems are now in evidence with considerable variation among the four texts.

In Grade 5 "total repetitions" vary from 34 to 75, which is just a little less than for the preceding grade. In Grade 6 there is a considerable falling off in Pattern A1 in all four texts; for the first time "multiple-step repetitions" slightly exceed "one-step repetitions" in Texts Y, Z, and W. The combined graph for all four grades explains itself.

The graphic chart of Pattern A2 is found on page 126. It will be remembered that Pattern A2 differs from Pattern A1 only in the omission of emphatic words, such as *both, in all, all together, together,* etc. In all other respects the two patterns are alike. In Grade 3 there is general agreement among Texts X, Z, and W with respect to the total offering of Pattern A2. In three texts this pattern appears exclusively in one-step problems.

In Grade 4 the total offering of Pattern A2 in Texts X, Y, and Z is larger than for Grade 3, whereas in Text W the total offering is smaller than for Grade 3. Pattern A2 appears in multiple-step problems in all four texts.

In Grade 5 Pattern A2 appears more frequently in multiple-step problems in Texts X and W than in one-step problems. One-step appearances of this pattern greatly exceed multiple-step appearances in Texts Y and Z. In Grade 6 there is closer agreement among all four texts with respect to total appearances than in any other grade.

The combined graph of all four grades evens up some of the discrepancies of the separate grades with respect to total offerings. These totals vary from 147 to 192. In the combined graph there is found more uniformity on the part of the four texts with respect to Pattern A2 than there is with respect to Pattern A1.

Addition Patterns A3 and A4

In the graphic representation of the findings on Pattern A3, given on page 127, it will be found that there is considerable variation in each grade in the offerings of the four authors. In Grade 3 there is wide variation among the authors in the total number of repetitions of this pattern; Author Y gives 28 repetitions, whereas Author X gives twice as many. In Grades 4, 5, and 6 the graphs show, in general, that Pattern A3 occurs much more frequently in multiple-step problems than in one-step problems.

In Grade 5 there is less variation in total repetitions and wide variation between the appearances in one-step problems and in multiple-step problems. In Grade 6 there is a marked falling off in Pattern A3 in all four texts. In the graph for all four grades combined the discrepancies in the separate grades are evened up with reference to "one-step repetitions." There is also less variation between the appearances of A3 in one-step problems as compared with multiple-step problems than was found in the separate grades.

From the graphic chart of Pattern A4, given on page 128, it is evident that this pattern receives very little attention in any of the four grades. Furthermore, the appearance of this pattern is very irregular. In Text Z it appears 20 times in Grade 3, 5 times in Grade 4, and once each in Grades 5 and 6. In Text W it appears 4 times in Grade 3, 7 times in Grade 4, 9 times in Grade 5, and 14 times in Grade 6. It appears seldom, if at all, in any grade in Texts X and Y. Total appearances, as shown in the graph for all four grades combined, vary from 5 to 34. In each case this is a very small offering spread over a period of four years.

It will be remembered that Pattern A4 is made up of those problems which are stated in much the same way as certain subtraction problems. This pattern is illustrated by the following problems: (1) "Mr. Smith has been offered a used automobile for $525. This, the owner says, is $320 *less than* the car cost when new. What was the cost of the car when new?" (2) "Tom had 15¢ left when he came home from the circus today. His trip to the circus, including carfare, cost him 60¢. How much money did he have when he left home for the circus?" In general, teachers say problems such as these are more difficult than problems of Patterns A1, A2, and A3. Some children confuse problems like the above with subtraction problems because they are influenced by words rather than the underlying thought of the problem. The criticism is sometimes raised that problems of Pattern A4, in general, are not real problems because the answer or sum has to be known before the problem can be stated. It is true, in many instances, that the answer is known to some person before the problem can be stated in the form under consideration, but the solution of the problem is often done by another person who does not know the answer. Problem (2), given above, could easily represent a real situation if Tom is not sure of the amount of money he had when he left home for the circus and if his mother is checking up on his bookkeeping. For example, Tom may have thought he had more than 75¢ when he left home; if that were the case, something is wrong since he spent only 60¢ and had 15¢ left when he came home. Problems of Pattern A4 often occur in life situations. Children need sufficient experience with such problems to enable them to recognize them as addition problems. In view of its difficulty it would seem reasonable to expect that Pattern A4 should receive considerably more emphasis than it has in the texts under discussion, especially in Grades 4, 5, and 6.

Comparisons and Relations Between Patterns

Table VII, shown below, gives for each of the texts the "total repetitions" of each addition pattern in all four grades combined. This table shows that Pattern A1 ranks first in three of the four texts and represents from 35.8% to 42.7% of the total offering in addition in these texts. At the other extreme, Pattern A4 ranks fourth in all four texts. Pattern A2 ranks second in two texts and third in the other two. Pattern A3 ranks first in one text, second in two texts, and third in one text. It is of

TABLE VII

SUMMARY OF THE TOTAL REPETITIONS OF EACH ADDITION PATTERN IN ALL FOUR GRADES COMBINED

Text	*Item*	*A1*	*A2*	*A3*	*A4*	*Total*
X	Number of repetitions	308	192	233	5	738
	Per cent	41.7	26.0	31.6	0.7	100
	Rank	1	3	2	4	
Y	Number of repetitions	246	175	147	8	576
	Per cent	42.7	30.4	25.5	1.4	100
	Rank	1	2	3	4	
Z	Number of repetitions	147	147	190	27	511
	Per cent	28.8	28.8	37.2	5.2	100
	Rank	2	2	1	4	
W	Number of repetitions	230	161	218	34	643
	Per cent	35.8	25.0	33.9	5.3	100
	Rank	1	3	2	4	

interest to observe that if the totals for Patterns A1 and A2 are combined for each series, there is complete agreement as to order of rank, with A1 and A2 combined ranking first, A3 ranking second, and A4, third.

The total repetitions of each pattern, as given in Table VII, are shown in graphic form in Chart 4, page 129. This chart makes it possible to compare the treatment of the four addition patterns in any one text with the corresponding treatment in each of the other texts.

For purposes of reference, Tables XIIIa, XIIIb, XIIIc, and XIIId are given in the Appendix. These tables summarize all the findings with respect to the four addition patterns for each of the four texts examined.

ADDITION PATTERNS

ADDITION PATTERN A1

Finding the Sum When the Problems Contain Such Words as *Both*, *Together*, *in All*

☐ Repetitions in one-step problems

■ Repetitions in multiple-step problems

Grade 3

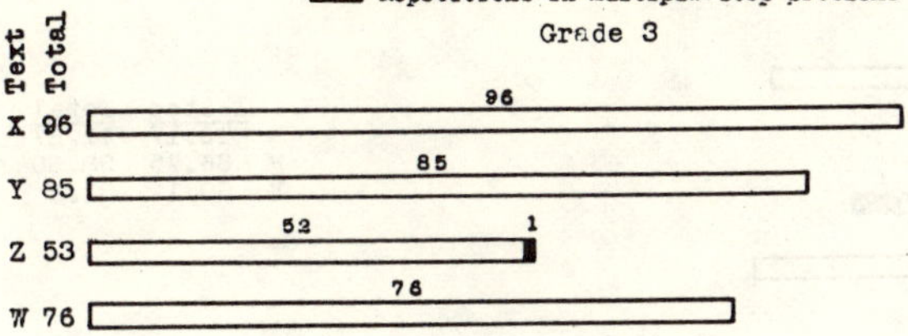

	1-Step	Total
σ	16.21	15.82
M	77.25	77.50
V	21.0	20.4

Grade 4

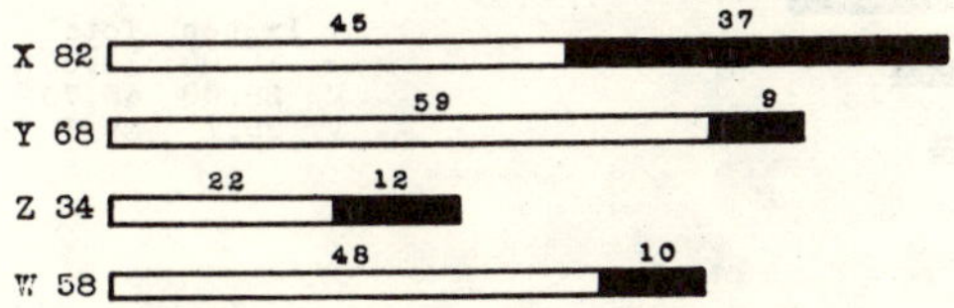

	1-Step	Total
σ	13.46	17.51
M	43.50	60.50
V	30.9	28.9

Grade 5

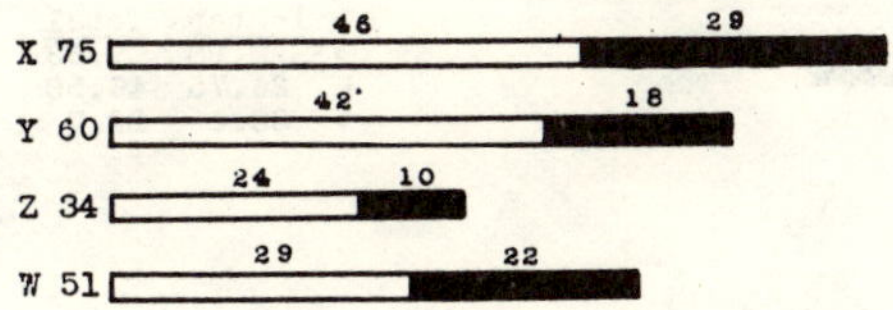

	1-Step	Total
σ	9.04	14.85
M	35.25	55.00
V	25.6	27.0

Grade 6

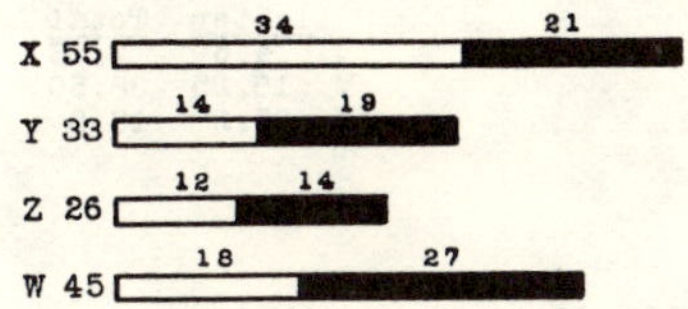

	1-Step	Total
σ	8.65	11.12
M	19.50	39.75
V	44.4	28.0

Grades 3 to 6 Combined*

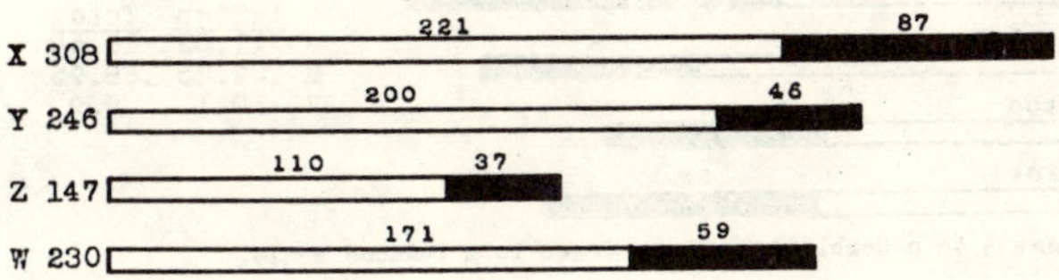

	1-Step	Total
σ	41.78	57.44
M	175.50	232.75
V	23.8	24.7

*The graph for Grades 3 to 6 Combined has been drawn to a reduced scale.

ADDITION PATTERN A2

Finding the Sum When the Problems Contain No Characteristic Words

Grade 3

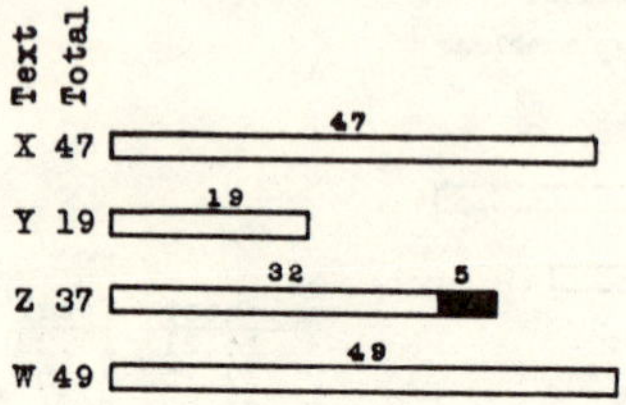

	1-Step	Total
σ	12.17	11.87
M	36.75	38.00
V	33.1	31.2

Grade 4

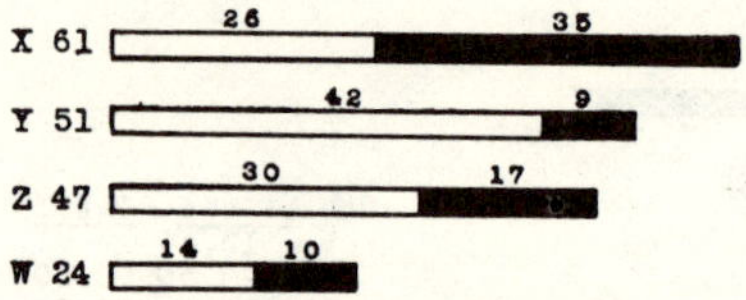

	1-Step	Total
σ	10.00	13.56
M	28.00	45.75
V	35.7	29.6

Grade 5

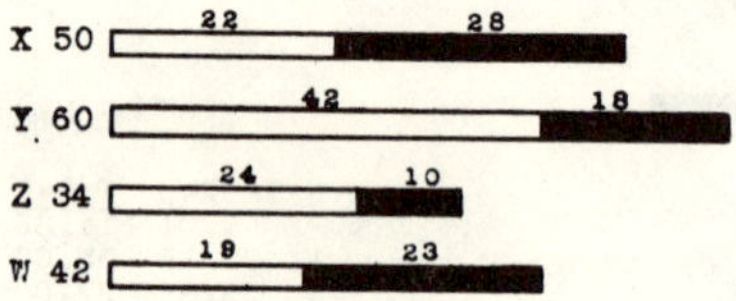

	1-Step	Total
σ	8.98	9.63
M	26.75	46.50
V	33.6	20.7

Grade 6

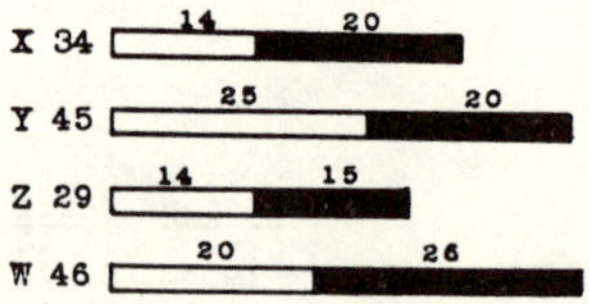

	1-Step	Total
σ	4.60	7.23
M	18.25	38.50
V	25.2	18.8

Grades 3 to 6 Combined*

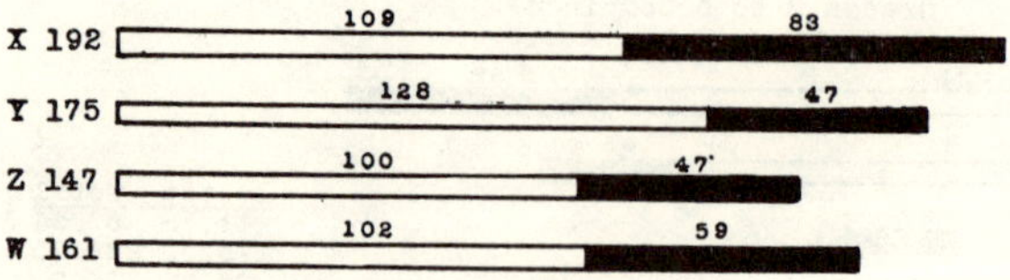

	1-Step	Total
σ	11.05	16.68
M	109.75	168.75
V	10.1	9.9

*The graph for Grades 3 to 6 Combined has been drawn to a reduced scale.

ADDITION PATTERN A3

Finding the Sum When the Problems Relate to Buying and Selling

Grade 3

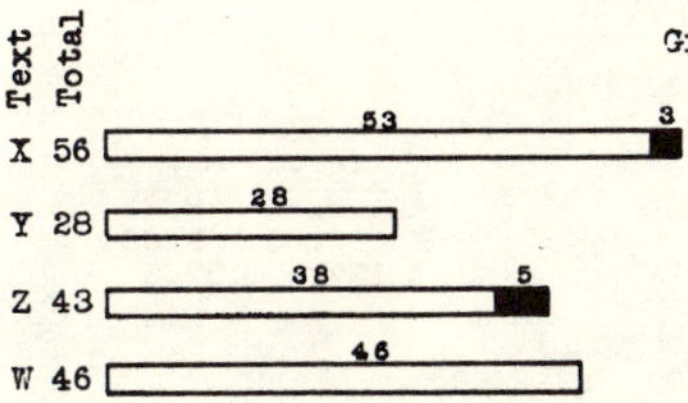

	1-Step	Total
σ	9.31	10.03
M	41.25	43.25
V	22.6	23.2

Grade 4

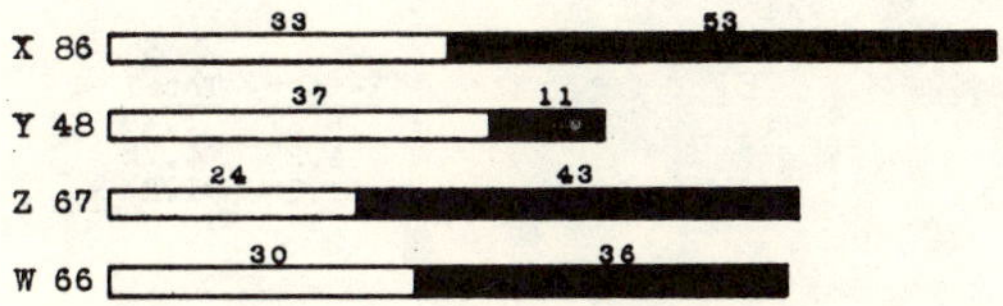

	1-Step	Total
σ	4.74	13.44
M	31.00	66.75
V	15.3	20.1

Grade 5

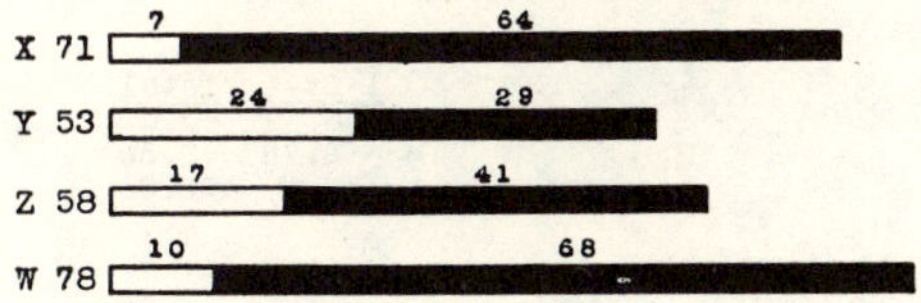

	1-Step	Total
σ	6.58	9.97
M	14.50	65.00
V	45.4	15.3

Grade 6

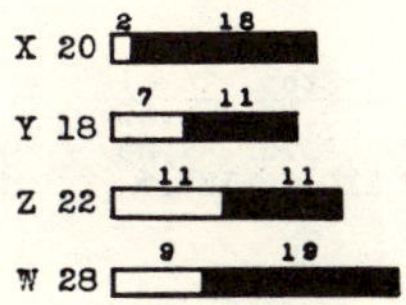

	1-Step	Total
σ	3.34	3.74
M	7.25	22.00
V	46.1	17.0

Grades 3 to 6 Combined*

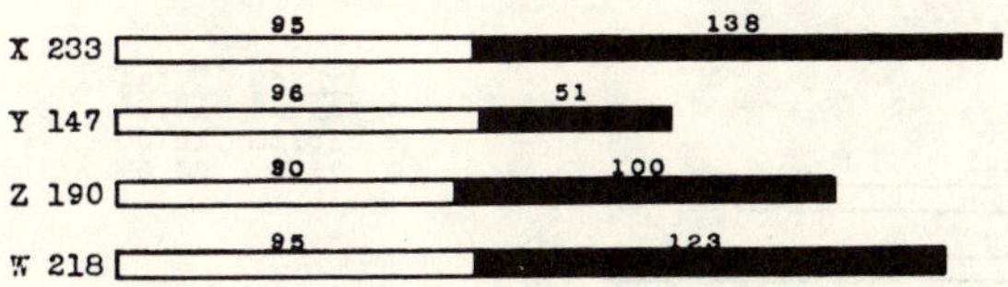

	1-Step	Total
σ	2.35	32.73
M	94.00	197.00
V	2.5	16.6

*The graph for Grades 3 to 6 Combined has been drawn to a reduced scale.

ADDITION PATTERN A4

Finding the Sum When the Problems Contain Phraseology Used in Subtraction

Grade 3

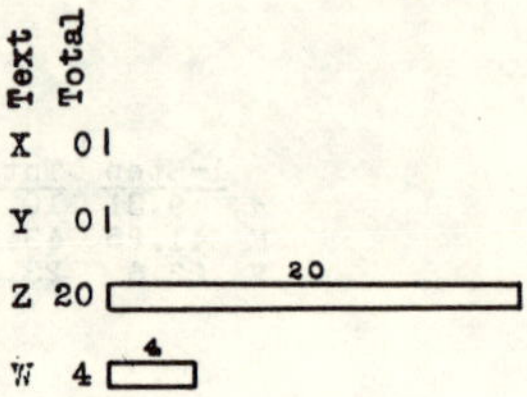

	1-Step	Total
σ	8.25	8.25
M	6.00	6.00
V	137.5	137.5

Grade 4

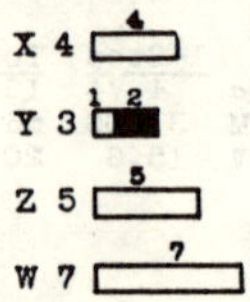

	1-Step	Total
σ	2.17	1.48
M	4.25	4.75
V	51.1	31.2

Grade 5

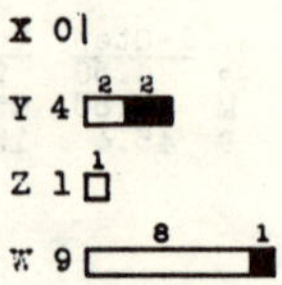

	1-Step	Total
σ	3.11	3.50
M	2.75	3.50
V	113.1	100.0

Grade 6

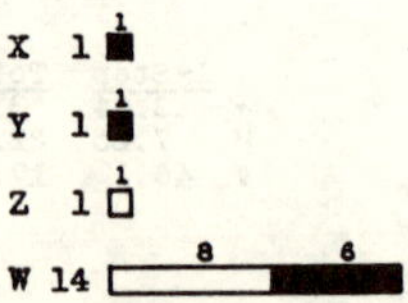

	1-Step	Total
σ	3.03	5.63
M	2.75	4.25
V	110.2	132.5

Grades 3 to 6 Combined

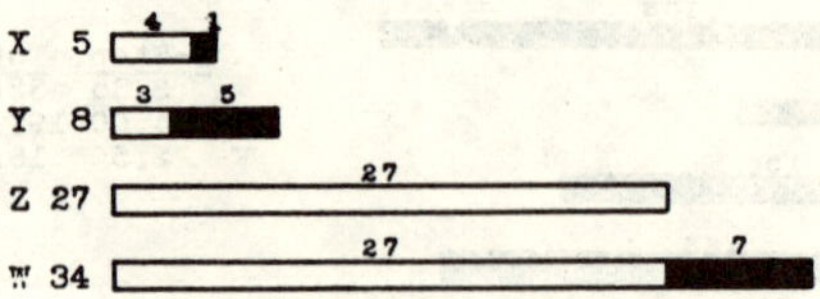

	1-Step	Total
σ	11.65	12.30
M	15.25	18.50
V	76.4	66.5

CHART 4

Total Repetitions of Each of the Addition Patterns
in All Four Grades Combined,
Arranged by Texts

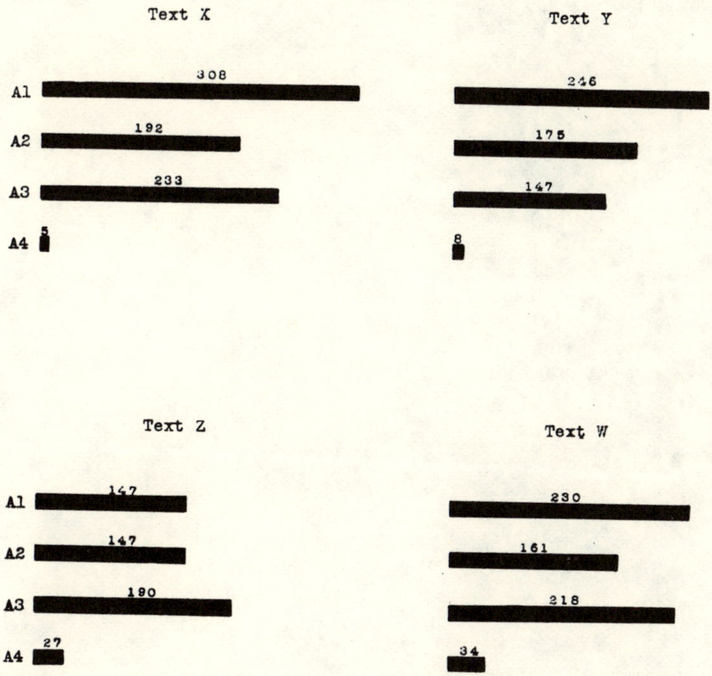

CHAPTER VII

COMPARISONS AND COMMENTS RELATING TO ALL THE PATTERNS

A COMPARISON OF THE FOUR TEXTS will now be made with respect to the relative emphasis given the four fundamental operations. Table VIII gives the total offering in all four grades of the patterns representing each operation.

TABLE VIII

TOTAL REPETITIONS OF THE PATTERNS IN THE FOUR OPERATIONS

	Text X	*Text Y*	*Text Z*	*Text W*	*Total*
Addition	738	576	511	643	2,468
Subtraction	774	796	684	744	2,998
Multiplication	947	1,070	1,121	1,221	4,359
Division	867	804	1,083	919	3,673
Total	3,326	3,246	3,399	3,527	13,498

In the first line of Table VIII, the total 738 for Text X represents the "total repetitions" for all four grades of all the addition patterns, in both one-step and multiple-step problems. The total 2,468 represents the "total repetitions" of the addition patterns for all four series combined.

Attention is called to the total offering of each series which is given in the last line of Table VIII. It is very significant that the total offering in pattern repetitions, representing all four operations, is so nearly alike for all four texts. These totals, in order of size, are 3,246 for Text Y, 3,326 for Text X, 3,399 for Text Z, and 3,527 for Text W. This means that between Text W, which has the largest total offering, and Text Y, which has the smallest, there is a difference of only 281 repetitions. This is a very small difference compared with the total offering of Text W. Text X has only 201 total repetitions less than Text W while Text Z has only 128 less than Text W. Thus, so far as their *total offering* in pattern repetitions is concerned, there is very close agreement among the four authors.

Table VIII also discloses two other important facts. If each operation is considered separately, closest relative agreement on total pattern repetitions exists in subtraction and least agreement exists in addition. In point of total emphasis, multiplication ranks first, division second, subtraction third, and addition fourth in all four series.

An examination of the offering in each operation will show in what respects there is agreement among the authors in the emphasis given from grade to grade. Table IX shows for each series the offering in pattern repetitions in addition.

TABLE IX

TOTAL REPETITIONS OF THE ADDITION PATTERNS BY GRADES

	Text X	*Text Y*	*Text Z*	*Text W*
Grade 3	199	132	153	175
Grade 4	233	170	153	155
Grade 5	196	177	127	180
Grade 6	110	97	78	133
Total	738	576	511	643

It is evident from this table that there is considerable variation among the texts in the offerings of a given grade. Texts X and Y have increased the offering in Grade 4 over that in Grade 3, as might be expected; in Text Z the offering in Grade 4 is the same as in Grade 3, while in Text W the fourth grade offering is less than that of the third grade. In Grade 5 it is seen that the offering in addition patterns is close to, or a little greater than, the corresponding offering in Grade 3, the reason for this being that problem solving in connection with the addition of fractions and decimals begins in Grade 5. All four authors have their smallest offering in Grade 6, which seems reasonable in view of the emphasis on the addition patterns given in the preceding grades.

A careful study of the subtraction offering in Table X reveals closer agreement among the authors than was found in addition.

TABLE X

TOTAL REPETITIONS OF THE SUBTRACTION PATTERNS BY GRADES

	Text X	*Text Y*	*Text Z*	*Text W*
Grade 3	222	160	232	180
Grade 4	225	184	140	163
Grade 5	192	253	182	238
Grade 6	135	199	130	163
Total	774	796	684	744

It is very significant that the total offering for all four grades in subtraction patterns varies so little from text to text. It is also significant that each text has a greater total offering in subtraction than in addition. Since the process of subtraction is more difficult than the process of addition, it is natural to expect a greater emphasis on the subtraction patterns than on the addition patterns. The rather liberal offering of the subtraction patterns in Grade 5 is explained by the fact that common fractions and decimals begin in that grade with their accompanying work in problem solving.

Before presenting the table which summarizes the total number of repetitions of the multiplication patterns in each grade, it will be helpful to recall an observation, pointed out in Chapter V, that all four texts have given an unusual amount of attention to Multiplication Patterns M1 and M2. In fact, these two patterns together represent from 79% to 84% of the total offering in multiplication in the four texts. It is the high frequency of these two patterns in each grade that raises the total offerings in each text to the large amounts shown in Table XI.

Table XI shows that all four authors give least emphasis to the multiplication patterns in Grade 3 and increase the emphasis in each succeeding grade over the previous grade. The largest offering occurs in Grade 6. There is considerable variation in the total offerings for all four grades. The total offering of Text W amounts to 1,221 repetitions, which is 30% greater than that found in Text X, which has a total of 947 repetitions.

TABLE XI

Total Repetitions of the Multiplication Patterns by Grades

	Text X	*Text Y*	*Text Z*	*Text W*
Grade 3	157	138	162	125
Grade 4	247	236	297	322
Grade 5	265	315	329	329
Grade 6	278	381	333	445
Total	947	1,070	1,121	1,221

In Grade 5 the very liberal offering in multiplication patterns in all four texts is due to the large representation of Pattern M2 in that grade. The still larger totals in Grade 6 are due not only to Pattern M2 but also to Pattern M7, which makes its first appearance in Grade 6. M7 is the pattern which is concerned with finding a per cent of a number.

Pattern M2, which is mentioned above, is the one that relates to finding the total cost when the number of like articles purchased and the price of each one are given. This pattern alone represents more than 50% of all the pattern repetitions in multiplication in each of the texts. Doubtless the very high frequency of Pattern M2 in all the texts is due to its wide social usefulness in buying and selling transactions. But granting the broad usefulness of this pattern, the question might well be raised whether the emphasis given this pattern is justifiable. It would seem better to give less emphasis to this relatively easy pattern and greater emphasis to certain division patterns which are more difficult.

The offering in division is summarized in Table XII which shows the repetitions in the division patterns for each series of texts.

Table XII explains itself so far as the first four lines are concerned. The figures in the last line of the table, however, deserve special study; these figures, which give the total offering in division for each text, may profitably be compared with the corresponding figures giving the total offerings in multiplication

TABLE XII

TOTAL REPETITIONS OF THE DIVISION PATTERNS BY GRADES

	Text X	*Text Y*	*Text Z*	*Text W*
Grade 3	126	99	157	140
Grade 4	287	193	328	272
Grade 5	250	243	316	261
Grade 6	204	269	282	246
Total	867	804	1,083	919

which are found in Table XI on page 134. Referring to these two tables, we find for Text X that division has 867 repetitions while multiplication has 947, or 9% more than division; for Text Y division has 804 repetitions while multiplication has 1,070, or 33% more than division; for Text Z division has 1,083 repetitions while multiplication has 1,121, or 4% more than division; for Text W division has 919 repetitions while multiplication has 1,221, or 33% more than division.

It is seen that in all four texts the multiplication patterns are more liberally represented than the division patterns. This arrangement seems unbalanced for two reasons: (1) division has 16 patterns while multiplication has only 8; (2) verbal problems in division involve more varied and difficult concepts, in general, than those of multiplication. In each of the four texts, the subtraction patterns, which are more difficult than the addition patterns, do receive more attention than the addition patterns. In other words, the subtraction patterns receive the attention one expects, while the division patterns do not. In view of these facts, there seems to be sound basis for stating that the division patterns should receive more attention than the multiplication patterns, especially in Grades 5 and 6.

Other Observations

A few comments will be made concerning the 38 patterns found in this study. Of the 38 patterns, 16 are related to division, which gives the operation of division the highest representation among all the patterns. In other words, the division patterns constitute about 42% of all the patterns. This large number of division patterns seems to be consistent with the difficulty of this topic as applied in problem solving.

Subtraction is allotted 10 patterns out of the 38, thus becoming next in importance to division; this allotment seems reasonable in view of the variety of concepts and phraseology represented by the subtraction patterns. Multiplication occupies third place with 8 patterns, while addition has only 4 patterns. The lesser attention given the addition and the multiplication patterns seems to agree with the experience of many teachers who have found that addition and multiplication as applied in problem solving are easier than subtraction and division.

Subtraction and division are both *inverse* processes, subtraction being the inverse of addition, and division the inverse of multiplication. Throughout arithmetic inverse operations and processes have uniformly caused more trouble to pupils than direct operations or processes. This situation is also found in percentage; Case I of percentage, which is the direct case, is the easiest; Cases II and III, which are the inverse cases, are much more difficult. The fact that division is represented by more patterns than addition and multiplication seems to imply that it takes a wider variety of quantitative experiences and problem situations to acquire the general concept of division than it does to acquire the general concepts of addition and multiplication.

Should Certain Patterns Be Omitted?

The question of whether certain of the 38 patterns should be omitted in Grades 3 to 6, or postponed to later grades, has in part been answered in the preceding chapters. The fact that certain patterns have had very little attention on the part of all four authors naturally raises such questions. It was found both in multiplication and in division that the patterns representing the concepts of distance, rate, and time had relatively small representation in any text and such representation as existed varied considerably from text to text. Concepts involving rate and time in relation to distance are difficult for pupils in the upper grades and high school as well as in the lower grades. It might be well seriously to consider these patterns as more suitable for the upper grades. This does not necessarily mean the total omission of these patterns, particularly in Grades 5 and 6; some of them might be retained for the abler pupils, or for the purpose of answering important quantitative questions that might arise in connection with pupil activities. Undoubtedly the patterns involving distance, rate, and time, both in multiplication and in division, will have increasing social application as the interest in air travel becomes more widespread.

Are More Patterns Needed?

Another question that might naturally arise is the following: Do 38 patterns completely cover the range of problem solving found in Grades 3 to 6? The answer is "no." Certain problems relating to mensuration, denominate numbers, and scale drawing were omitted in this study in order to limit its scope. There would have to be, therefore, a few patterns to cover these situations. Further, some new patterns would also arise in Grades 7 and 8 where provision would have to be made for Case II and Case III of percentage. No analysis of a large body of problem material has been made to determine definitely how many patterns would be involved of the kinds just mentioned. Based upon the experience gained in collecting the 38 patterns, it seems safe to state, however, that the total number of patterns, covering the additional needs outlined throughout Grade 8, would be less than 50, these 50 including the 38 patterns already found.

If some other investigator should independently make the analysis of the four textbooks used in this study, it is possible that he might arrive at an arrangement and number of patterns differing somewhat from the one given here. If that should happen, it does not seem serious. If the work were carefully done, and if the number of patterns were not too few, it would probably serve the same purpose as the patterns in this study, namely, that of building up generalizations or concepts concerning the processes of addition, subtraction, multiplication, and division through acquaintance with detailed aspects of those concepts as represented by the patterns.

Do the Patterns Cover All Aspects of Problem Solving?

It is not claimed that the 38 patterns discussed in this study are a total solution to the broad problem of teaching problem solving. They seem to offer a solution so far as one-step problems are concerned since every one-step problem can be handled in terms of some one of the 38 patterns. With respect to multiple-step problems, the patterns cover each of the individual steps of these problems but they do not provide for the sequence of steps. For example, if a boy has 75¢ and wants to know how much he will have left if he buys a top for 15¢ and a kite for 25¢, the patterns provide for the *addition step* in which 15¢ and 25¢ are added and also for the *subtraction step* in which 40¢ is subtracted from 75¢. But the patterns do not cover that phase of the thinking which suggests that one *adds first* and *subtracts second*. Multiple-step problems, like the one just mentioned, occur frequently in connection with retail buying where several articles are purchased and the total cost is subtracted from a given amount of money. In fact, these particular multiple-step problems represent a type in themselves; after doing a certain number of problems of this type, one automatically *adds the purchases* and *subtracts the result from the given amount of money*. In other words, after repeated experiences with this type of problem, the order of operations requires no special thinking on the part of the pupil. The situation is similar where one finds the average of several numbers, which is a two-step problem. By repeated experience with such problems one automatically *adds the numbers first* and *then divides the result by the number of items.* Thus the order of steps in this type of problem soon becomes automatic. A third illustration is related to discount problems, such as the case where one wants to know what he will have to pay for an article that has a marked price of $1.50 and is to be sold at a sale for "1/3 off." There are many discount problems of this type. In these problems, most people *first find the discount* and *then subtract the result from the marked price.* In other words, the sequence of operations becomes automatic. A number of other illustrations of problems of frequent occurrence, where the sequence of steps becomes automatic, could be given. In these problems the sequence of steps forms a pattern, which might be designated a *sequence pattern,* to distinguish it from the 38 *thought patterns* found in this study. After the pupils have been given experience in solving problems of the types that form *sequence patterns,* it is assumed that they will have no further difficulty with the *order of steps* in working such problems. An interesting task would be the investigation of those multiple-step problems that conform to the sequence patterns.

Summary

The 38 thought patterns provide for all one-step problems; together with the sequence patterns, they also provide for a large group of multiple-step problems. There still remains, however, that group of multiple-step problems which do not fall into sequence patterns simply because they do not occur sufficiently often to impress the sequence upon one's mind. It is these problems that present a real challenge to the teacher. Unfortunately there has been no wide investigation to indicate how pupils can most successfully attack the solution of such problems. Of the various attempts that have been made in this direction, perhaps the most promising is the one first made by Clark and Vincent,[1] and later repeated more extensively by Hanna[2]; each of these three investigators teaches the pupil how to attack the solution of problems by a method of analysis, called the "dependencies method," which is similar to that used in solving original exercises in geometry. If further experimentation should show the way to make the method of analysis usable in schools in general, then that method, coupled with the thought patterns and the sequence patterns already mentioned, should help materially to improve the teaching of problem solving in elementary arithmetic.

[1] J. R. Clark and E. L. Vincent, "A Comparison of Two Methods of Arithmetic Problem Analysis," *Mathematics Teacher,* 18: 226–233, 1925.
[2] P. R. Hanna, *Arithmetic Problem Solving.* 1929.

CHAPTER VIII

FINDINGS AND IMPLICATIONS

Findings

THE FINDINGS OF THIS STUDY may be divided into two groups as follows: (1) findings relating to the discovery of the number of one-step patterns that are represented in the verbal problems collected from four series of arithmetics covering Grades 3 to 6 inclusive and analyzed in this study, together with the description and classification of these patterns; and (2) findings relating to the frequency and distribution of these patterns in the four series of arithmetics.

FINDINGS CONCERNING THE PATTERNS THEMSELVES

1. A critical analysis of the one-step, two-step, and three-step verbal problems found in the four series of arithmetics, covering Grades 3 to 6 inclusive, shows that these problems can be broken down into 38 one-step patterns. Of these one-step patterns, 4 are in addition, 10 in subtraction, 8 in multiplication, and 16 in division. The four series of arithmetics contain a total of about 15,000 verbal problems, all of which were classified except a few types of problems that were omitted to limit the scope of this investigation. The fact that 38 patterns are sufficient to cover such a large number of problems is the most important finding in this study.

2. The underlying thought in all 4 addition patterns is that of combining two or more numbers of the same kind to form a whole.

3. Some of the 10 subtraction patterns are related to each other in their basic meaning and can be arranged in four major groups as follows:

GROUP I. Three patterns that involve the remainder idea.
GROUP II. One pattern that involves the building-up idea.
GROUP III. Five patterns that involve the comparison idea.
GROUP IV. One pattern that involves the separation-into-parts idea.

4. Some of the 8 multiplication patterns are so related in their underlying thought that they can be arranged in five major groups as follows:

GROUP I. Two patterns that require finding the total amount, exclusive of distance.
GROUP II. Two patterns requiring the total distance and involving the concepts of rate and time.
GROUP III. Two patterns requiring a fractional part of a number to be found.
GROUP IV. One pattern requiring a per cent of a number to be found.
GROUP V. One pattern requiring a number to be found when a part of it is given.

5. The division patterns are equally divided between the two notions of division; 8 patterns represent the measurement concept of division, and 8, the partition concept of division. Certain relationships existing among these patterns make it possible to group them into six major groups, three of which represent the measurement concept and three the partition concept.

The three major groups representing the measurement concept of division are as follows:

GROUP I. Three simpler patterns requiring that a number of groups be found.
GROUP II. Three patterns involving concepts of distance, rate and time.
GROUP III. Two patterns involving the comparison of two numbers.

The three major groups representing the partition concept of division are as follows:

GROUP IV. Four patterns that involve finding the amount, size, or cost of each part.

GROUP V. One pattern that involves finding an average but not related to distance, rate, and time.

GROUP VI. Three patterns that involve finding an average in relation to distance, rate, and time.

Findings Concerning the Frequency of the Patterns in Four Series of Texts

1. The addition patterns A1, A2, and A3 occur with much greater frequency than A4. In three of the four texts, Pattern A1 appears with greatest frequency, whereas A3 appears with greatest frequency in the fourth text. A2 holds second rank in two texts and third rank in two texts. A4 appears infrequently in all four texts.

2. Of the four ideas of subtraction, the remainder and the comparison ideas receive greater emphasis than the building-up and separation-into-parts ideas. In two series the comparison idea receives considerably more emphasis than the remainder idea, whereas the remainder idea receives more emphasis than the comparison idea in the other two series. The building-up idea receives a little more emphasis than the separation-into-parts idea in three of the series.

3. The subtraction patterns which appear with greatest frequency in all four texts are S1 (finding how much is left) and S5 (comparison using *how much more* or *less*).

4. Of the five major groups of multiplication patterns, Group I (finding the total amount) ranks first in frequency in all four series.

5. Multiplication patterns which appear with greatest frequency in all four series are M2 (finding total cost) and M1 (finding total amount).

6. In three series, the patterns representing the partition concept of division considered as a whole appear with greater frequency, whereas in the fourth series the patterns representing the measurement notion of division appear with greater frequency.

7. Of the three major groups of division patterns representing the measurement concept, Group I (covering the simpler patterns) ranks first; this group constitutes about one third of the entire offering in division for each series. Of the three groups representing the partition concept, Group IV (finding the size of each part) ranks first in each text and also constitutes about one third of the entire offering in division for each series.

8. Of the 8 division patterns in the measurement group, Pattern D1 (finding the number of groups) ranks highest. Of the 8 patterns in the partition group, there is no uniformity among the four series with respect to the pattern that ranks highest; Patterns D9 and D10, however, are consistently high in all four series if taken together as one unit.

9. There is, in general, wide variation among the four authors with respect to the "total repetitions" of each pattern in each grade. Coefficients of variation exceeding 40 are not uncommon.

10. There is still wider variation, in general, with respect to the "one-step repetitions" of each pattern in each grade. Coefficients of variation of 50 or more are not infrequent.

11. The "total repetitions" and the "one-step repetitions" of each pattern with respect to all four grades combined usually show less variation than is found in the individual grades.

12. Out of the 38 patterns, 28 patterns are represented in the third grade. Practically all of these 28 patterns are represented exclusively in one-step problems in this grade. Repetitions in multiple-step problems are negligible in Grade 3.

13. In Grades 4 to 6, repetitions of the patterns in multiple-step form occur least frequently in the division patterns. This means that the division patterns in all grades have, in general, wider representation in one-step form than in multiple-step form. Repetitions in multiple-step form occur more frequently in addition and multiplication than in subtraction.

14. If one considers the *total offering of patterns covering all four operations in all four grades,* one finds that these totals are *nearly alike* for all four texts, varying from 3,246 repetitions to 3,527 repetitions. There is, therefore, very close agree-

ment among the four authors with respect to their total offering of all patterns in all four years.

15. If one considers the *total treatment of the patterns of each operation* for all four grades, one finds in all four series that the multiplication patterns rank first in attention, division second, subtraction third, and addition fourth. In all four texts certain multiplication patterns seem to have excessive treatment at the expense of the division patterns.

Implications

The one-step problem patterns discussed in this study are attempting to do for problem solving in arithmetic what Thorndike began to do about twenty-five years ago with respect to the four operations with abstract numbers. Through the efforts of Thorndike and later investigators analyses were made of the successive difficulties that arise as a pupil is learning each of the four fundamental operations in arithmetic. Thorndike[1] also analyzed the abstract exercises in textbooks in arithmetic that were in use at that time and found many shortcomings in these texts in respect to the grading of the abstract exercises. There was overemphasis of one difficulty at the expense of another. Some types of difficulty were represented not at all or so meagerly as to be of little service. The distribution of the exercises of each type over the several grades appeared to follow no systematic plan.

During the past fifteen years the grading of abstract exercises to see that each type of difficulty is adequately represented has been given considerable attention by the authors of most of the textbooks in arithmetic in use today. The authors of these books have definitely planned the abstract exercises so that each type of difficulty is adequately represented in each grade. It is a rather easy matter to criticize a modern textbook if it is deficient in any of these details that relate to the process of

[1] E. L. Thorndike, *The Psychology of Arithmetic.* 1922.

computing. But such criticism is possible today only because Thorndike and others pointed out the necessity of taking each of these types of difficulty into consideration.

The one-step problem patterns in this study represent an attempt to do for the solution of verbal problems in arithmetic what the work by Thorndike and others has done to improve the teaching of computation. This may be illustrated by referring to the problem patterns for subtraction. If the pupil is given a variety of one-step problems belonging to Pattern S1, which is concerned with finding how much is left, these particular one-step problems contribute to the formulation of a generalization about Pattern S1. In like manner, if the pupil does a number of problems belonging to Pattern S5, which is concerned with the comparison of two numbers in using the phrases *how much more* or *how much less,* he comes to formulate a generalization about the comparison of numbers in such cases. Each pattern, in turn, contributes to the still larger generalizations regarding the process of subtraction as a whole. Problems involving each pattern broaden the pupil's concept of subtraction until finally subtraction comes to mean a process which is used (1) when a given amount has been decreased and the amount left is required, (2) when a given amount is on hand and it is required to find how much more is needed to equal a desired larger amount, (3) when two numbers are to be compared by finding their difference, and (4) when the whole and one part are given and it is required to find the other part. Generally, wide experience with many problems of each pattern, with the fundamental thought expressed in a variety of ways, is necessary in order to acquire the broad generalization relating to subtraction. In a number of instances these problem patterns represent types of quantitative thinking associated with certain life experiences. To see that all these patterns are well represented in elementary school arithmetic is also to assure that the pupil comes in contact with a large variety of life experiences involving quantitative thinking, such as buying, selling, sharing, comparing, and traveling.

Use of the Patterns by Teachers and Others

The patterns are not for the pupils to use as such. In general, most pupils will become familiar with the type of thinking involved in a pattern by solving a variety of one-step problems representing that pattern. The patterns will be of service to the teacher, however, to make sure that important types of quantitative thinking and the experiences related to them are not neglected in the class in arithmetic. Since patterns are not for pupils, it is not serious if two teachers should disagree upon the pattern represented by a given problem. If one teacher places the problem under one pattern and another places it under a second pattern, it is probable that one of these two patterns can be looked upon as a little more general form of the other pattern. The patterns are not necessarily independent of each other. The patterns do serve the practical purpose, however, of seeing that important types of thinking are not neglected.

The problem patterns of this study will probably be useful to the following groups of persons: (1) teachers and supervisors of arithmetic; (2) instructors in methods in arithmetic in teacher-training institutions; (3) authors of textbooks in arithmetic; (4) authors of problem-solving tests or achievement tests in arithmetic; (5) the makers of curricula in arithmetic.

One possible use of these problem patterns has been illustrated in this study in connection with the analysis of the problem material in textbooks to see to what extent the various types of problems are represented. Another possible use of these problem patterns would be in criticizing a test in problem solving in arithmetic or in presenting norms for constructing such a test. The writer has analyzed the problems in the problem-solving portion of a well-known achievement test in arithmetic, covering Grades 4 to 6, to see to what extent the 38 patterns are represented in that test. The results show that 19, or exactly half, of the 38 patterns are represented. The 19 patterns omitted include 6 patterns involving the distance-rate-time concepts which are difficult and which may properly be omitted in a test of this kind. Pattern M7 (percentage) is also omitted, which seems proper in a test intended for wide use in the intermediate grades. This makes 7 patterns whose omission is not questioned. Deducting these 7 omissions, there are still 12 patterns that are not represented: A4, S2, S3, S4, S6, S9, S10, D2, D3, D7, D9, and D12. It is interesting that so many of these omitted patterns represent the inverse operations of subtraction and division.

The Question of Transfer

An experiment in problem solving in arithmetic conducted by Osburn and Drennan[2] gives encouragement that the use of patterns like those in this study will bring about a real improvement in the ability of pupils to solve problems. This experiment was conducted in the third grade. Pupils were taught how to solve a small group of problems relating to each of the four operations. The problems were simply expressed and covered different types of quantitative situations. To gain experience with each type of problem, the pupil solved many similar problems. After six weeks of such instruction, confined to these groups of problems, a test was given which was made up of new problems presenting the same situations as the original problems but expressed in language differing from that in the original problems. The results of this test showed that there was considerable transfer of ability to solve these new problems, even in the cases of pupils of low intelligence. The results were so favorable as to lead the authors to say, "These results indicate that we are at last coming upon a sound technique for problem solving. Teach a few of the most important problem types thoroughly and depend upon transfer for the remainder."[3]

The problems used in the above experiment correspond to certain of the problem patterns in the present study. The favorable results obtained by Osburn and Drennan encourage the writer to believe that if a pupil has sufficient experience with a variety of problems representing a given problem pat-

[2] W. J. Osburn and L. J. Drennan, "Problem Solving in Arithmetic," *Educational Research Bulletin of Ohio State University,* 10: 123–128, 1931.

[3] *Ibid.,* p. 128.

tern, his ability to solve problems of that pattern will transfer to other problems of the same pattern.

The Method of Teaching Abstract Subtraction

On page 43 of this study the subtraction patterns represented in the four series of arithmetics examined by the writer are grouped according to the four important concepts of subtraction. These concepts are the remainder idea, the building-up idea, the comparison idea, and the separation-into-parts idea. Of these four ideas, the remainder idea, which answers the question *how much is left?*, occupies one third the attention of Authors X and Y and a little more than one third for Authors Z and W. The rest of the attention is given to the other three ideas of subtraction, none of which is concerned with answering the question *how much is left?*

It seems to the writer that this finding has a partial bearing upon a discussion that occurs frequently in school systems with respect to methods of teaching abstract subtraction. There are three methods of subtracting abstract numbers that are in common use in our elementary schools, namely, the take-away method, the additive method, and the equal-additions method. Many teachers argue that the take-away method of abstract subtraction is the one that should be employed since it represents the take-away idea which, they usually claim, is the one most frequently met in solving verbal problems in arithmetic. In fact, there are a number of teachers who have never realized that subtraction problems can represent anything but take-away situations. It will be of interest to these teachers to learn that of the four texts examined only one third, or a little more, of all the subtraction problems involve the take-away idea. Since the take-away method of abstract subtraction is often selected because of the belief that take-away situations outweigh all others in subtraction, it seems important to point out that this false assumption should not be the sole basis for adopting the take-away method of subtracting whole numbers.

Any particular school system has to consider many questions in selecting the method of abstract subtraction to be taught.

These usually include such matters as the personal experience of the teachers with these methods, the relative ease of explaining the three methods, the reactions of pupils, the question of switching from one method to another, the desirability of having only one method taught in a given school system, and the results of such scientific investigations as are available concerning this controversy. J. T. Johnson[4] has carried on some carefully controlled experimentation with reference to this question in which the three methods of subtraction have been compared with respect to accuracy of results and speed in computing. His findings deserve careful consideration whenever the discussion of this topic arises.

[4] J. T. Johnson, *The Relative Merits of Three Methods of Subtraction.* 1938.

BIBLIOGRAPHY

BODE, BOYD H. *How We Learn.* D. C. Heath and Company, Boston, 1940.

BOND, ELIAS A. *The Professional Treatment of the Subject Matter of Arithmetic, Grades One to Six.* Contributions to Education, No. 525. Bureau of Publications, Teachers College, Columbia University, New York, 1934.

BUSWELL, GUY T. "Summary of Arithmetic Investigations." *Elementary School Journal,* Vol. 28, pp. 702–709, 730–742, May, 1927; Vol. 29, pp. 691–698, 737–747, May, 1928; Vol. 30, pp. 766–775, May, 1929.

BUSWELL, GUY T. and JUDD, CHARLES H. *Summary of Educational Investigations Relating to Arithmetic.* Supplementary Educational Monograph, No. 27. The University of Chicago, June, 1925.

BROWNELL, WILLIAM A. and STRETCH, L. B. *The Effect of Unfamiliar Settings on Problem Solving.* Duke University Research Studies in Education, No. 1. Duke University Press, Durham, N. C., 1931.

BRUECKNER, LEO J. *Diagnostic and Remedial Teaching in Arithmetic.* The John C. Winston Company, Philadelphia, 1930.

CLARK, JOHN R. and VINCENT, LEONA E. "A Comparison of Two Methods of Arithmetic Problem Analysis." *Mathematics Teacher,* Vol. 18, pp. 226–233, April, 1925.

COMMINS, WILLIAM D. *Principles of Educational Psychology.* The Ronald Press Company, New York, 1937.

DURELL, THOMAS J. "Problem Solving in Arithmetic." *1941 Yearbook of Association of Mathematics Teachers of New Jersey,* pp. 34–38, December, 1941.

FORAN, THOMAS G. "Methods of Teaching Problem Solving in Arithmetic." *Catholic Education Review,* Vol. 32, pp. 269–282, May, 1934.

GROSSNICKLE, FOSTER E. "Cues in Division Problems Given in Nine Representative Textbooks in Arithmetic." *Elementary School Journal,* Vol. 33, pp. 452–461, February, 1933.

HANNA, PAUL R. *Arithmetic Problem Solving: A Study of the Relative Effectiveness of Three Methods of Problem Solving.* Bureau of Publications, Teachers College, Columbia University, New York, 1929.

JOHNSON, JOHN T. *The Relative Merits of Three Methods of Subtraction.* Contributions to Education, No. 738. Bureau of Publications, Teachers College, Columbia University, New York, 1938.

KRAMER, GRACE A. *The Effect of Certain Factors in the Verbal Arithmetic Problem upon Children's Success in the Solution.* Johns Hopkins University Studies in Education, No. 20. Johns Hopkins Press, Baltimore, 1933.

LENNON, ROGER T. *The Effect of Verbal Cues and Irrelevant Details in the Statement of Arithmetical Problems upon Children's Attempted Solutions.* Unpublished Master's Thesis, Fordham University, New York, 1937.

LUTES, OLIN S. *An Evaluation of Three Techniques for Improving Ability to Solve Arithmetic Problems.* University of Iowa Monographs in Education, First Series, No. 6. University of Iowa, Iowa City, 1926.

McEWEN, NOBLE R. *The Effect of Selected Cues in Children's Solutions of Verbal Problems in Arithmetic.* Duke University Research Studies in Education. Duke University Press, Durham, N. C., 1941.

MORTON, ROBERT L. *Teaching Arithmetic in the Elementary School.* Vol. 1, Primary Grades; Vol. 2, Intermediate Grades. Silver Burdett Company, New York, 1937, 1938.

MOSSMAN, EDITH L. "The Primary Purpose of Training in Mathematics—Not a Kit of Tools But a Way of Thinking." *School Science and Mathematics,* Vol. 38, pp. 992–1002, December, 1938.

NATIONAL COUNCIL OF TEACHERS OF MATHEMATICS. *The Teaching of Arithmetic.* Tenth Yearbook (William D. Reeve, edr.). Bureau of Publications, Teachers College, Columbia University, New York, 1935.

NATIONAL COUNCIL OF TEACHERS OF MATHEMATICS. *Arithmetic in General Education.* Sixteenth Yearbook (William D. Reeve, edr.). Bureau of Publications, Teachers College, Columbia University, New York, 1941.

NEWCOMB, RALPH S. *Modern Methods of Teaching Arithmetic.* Houghton Mifflin Company, Boston, 1926.

OSBURN, WORTH J. *Corrective Arithmetic, Vol. 2.* Houghton Mifflin Company, Boston, 1929.

OSBURN, WORTH J. and DRENNAN, L. J. "Problem Solving in Arithmetic." *Ohio State University, Educational Research Bulletin,* Vol. 10, pp. 123–128, March, 1931.

STONE, JOHN C. *The Teaching of Arithmetic.* Benjamin H. Sanborn and Company, Chicago, 1919.

THORNDIKE, EDWARD L. *The New Methods in Arithmetic.* Rand McNally and Company, Chicago, 1921.

THORNDIKE, EDWARD L. *Psychology of Arithmetic.* The Macmillan Company, New York, 1922.

WELCH, GERTRUDE L. and SHIBLES, MARK R. "A Fifth Grade Informal Problem Unit in Arithmetic on the Ordering and Distributing of Milk in the School." *Education,* Vol. 61, pp. 458–463, April, 1941.

WHITE, HELEN M. "Does Experience in the Situation Involved Affect the Solving of a Problem?" *Education,* Vol. 54, pp. 451–455, April, 1934.

WILSON, GUY M., STONE, M. B., and DALRYMPLE, C. O. *Teaching the New Arithmetic.* McGraw-Hill Book Company, Inc., New York, 1939.

APPENDIX

TABLES GIVING RESULTS OF ANALYSIS OF TEXTS X, Y, Z, AND W

TABLE XIIIa --- TEXT X

Summary of Addition Patterns Found in the One-Step and Multiple-Step Verbal Problems of Text X

The numbers in the line marked "1-step" show the repetitions of the pattern found in one-step problems. The numbers in the line marked "Hidden-step" show the repetitions of the pattern found in the hidden steps of multiple-step problems; the numbers in the line marked "Final step" give the repetitions found in the final steps of the multiple-step problems.

	Type of Repetition	Addition Patterns				
		A1	A2	A3	A4	Total
Grade 3	1-step	96	47	53	0	196
	Hidden-step	0	0	1	0	1
	Final Step	0	0	2	0	2
	Total	96	47	56	0	199
Grade 4	1-step	45	26	33	4	108
	Hidden-step	30	29	33	0	92
	Final Step	7	6	20	0	33
	Total	82	61	86	4	233
Grade 5	1-step	46	22	7	0	75
	Hidden-step	29	26	39	0	94
	Final Step	0	2	25	0	27
	Total	75	50	71	0	196
Grade 6	1-step	34	14	2	0	50
	Hidden-step	20	19	13	0	52
	Final Step	1	1	5	1	8
	Total	55	34	20	1	110

Summary for Grades 3 to 6 Inclusive

	A1	A2	A3	A4	Total
Total	308	192	233	5	738
%	41.7	26.0	31.6	0.7	100
Rank	1	3	2	4	

TABLE XIIIb --- TEXT Y

Summary of Addition Patterns Found in the One-Step and Multiple-Step Verbal Problems of Text Y

	Type of Repetition	Addition Patterns				
		A1	A2	A3	A4	Total
Grade 3	1-step	85	19	28	0	132
	Hidden-step	0	0	0	0	0
	Final Step	0	0	0	0	0
	Total	85	19	28	0	132
Grade 4	1-step	59	42	37	1	139
	Hidden-step	7	6	6	2	21
	Final Step	2	3	5	0	10
	Total	68	51	48	3	170
Grade 5	1-step	42	42	24	2	110
	Hidden-step	14	14	8	2	38
	Final Step	4	4	21	0	29
	Total	60	60	53	4	177
Grade 6	1-step	14	25	7	0	46
	Hidden-step	17	16	4	0	37
	Final Step	2	4	7	1	14
	Total	33	45	18	1	97

Summary for Grades 3 to 6 Inclusive

	A1	A2	A3	A4	Total
Total	246	175	147	8	576
%	42.7	30.4	25.5	1.4	100
Rank	1	2	3	4	

TABLE XIIIc --- TEXT Z

Summary of Addition Patterns Found in the One-Step and Multiple-Step Verbal Problems of Text Z

	Type of Repetition	Addition Patterns				
		A1	A2	A3	A4	Total
Grade 3	1-step Hidden-step Final Step	52 0 1	32 0 5	38 1 4	20 0 0	142 1 10
	Total	53	37	43	20	153
Grade 4	1-step Hidden-step Final Step	22 10 2	30 11 6	24 12 31	5 0 0	81 33 39
	Total	34	47	67	5	153
Grade 5	1-step Hidden-step Final Step	24 8 2	24 8 2	17 17 24	1 0 0	66 33 28
	Total	34	34	58	1	127
Grade 6	1-step Hidden-step Final Step	12 13 1	14 13 2	11 5 6	1 0 0	38 31 9
	Total	26	29	22	1	78

Summary for Grades 3 to 6 Inclusive

	A1	A2	A3	A4	Total
Total	147	147	190	27	511
%	28.8	28.8	37.2	5.2	100
Rank	2	2	1	4	

TABLE XIIId --- TEXT W

Summary of Addition Patterns Found in the One-Step and Multiple-Step Verbal Problems of Text W

	Type of Repetition	Addition Patterns				
		A1	A2	A3	A4	Total
Grade 3	1-step Hidden-step Final	76 0 0	49 0 0	46 0 0	4 0 0	175 0 0
	Total	76	49	46	4	175
Grade 4	1-step Hidden-step Final Step	48 8 2	14 8 2	30 11 25	7 0 0	99 27 29
	Total	58	24	66	7	155
Grade 5	1-step Hidden-step Final Step	29 21 1	19 20 3	10 37 31	8 0 1	66 78 36
	Total	51	42	78	9	180
Grade 6	1-step Hidden-step Final Step	18 24 3	20 25 1	9 9 10	8 0 6	55 58 20
	Total	45	46	28	14	133

Summary for Grades 3 to 6 Inclusive

	A1	A2	A3	A4	Total
Total	230	161	218	34	643
%	35.8	25.0	33.9	5.3	100
Rank	1	3	2	4	

TABLE XIVa --- TEXT X

Summary of Subtraction Patterns Found in the One-Step and Multiple-Step Verbal Problems of Text X

	Type of Repetition	Subtraction Patterns										
		S1	S2	S3	S4	S5	S6	S7	S8	S9	S10	Total
Grade 3	1-step	63	3	6	44	45	9	24	7	0	19	220
	Hidden-step	0	0	0	0	0	0	0	0	0	0	0
	Final Step	0	2	0	0	0	0	0	0	0	0	2
	Total	63	5	6	44	45	9	24	7	0	19	222
Grade 4	1-step	9	3	2	7	37	23	41	10	0	12	144
	Hidden-step	7	1	2	1	0	1	0	0	0	1	13
	Final Step	23	24	2	5	2	1	0	8	0	3	68
	Total	39	28	6	13	39	25	41	18	0	16	225
Grade 5	1-step	11	2	1	2	22	14	18	20	1	3	94
	Hidden-step	10	0	0	0	0	0	0	4	0	1	15
	Final Step	28	25	1	3	9	3	0	13	0	1	83
	Total	49	27	2	5	31	17	18	37	1	5	192
Grade 6	1-step	11	1	1	3	22	10	5	14	1	7	75
	Hidden-step	6	0	1	2	0	3	0	6	0	0	18
	Final Step	17	5	1	4	2	0	1	5	1	6	42
	Total	34	6	3	9	24	13	6	25	2	13	135

Summary for Grades 3 to 6 Inclusive

	S1	S2	S3	S4	S5	S6	S7	S8	S9	S10	Total
Total	185	66	17	71	139	64	89	87	3	53	774
%	23.9	8.5	2.2	9.2	18.0	8.3	11.5	11.2	0.4	6.8	100
Rank	1	6	9	5	2	7	3	4	10	8	

TABLE XIVb --- TEXT Y

Summary of Subtraction Patterns Found in the One-Step and Multiple-Step Verbal Problems of Text Y

	Type of Repetition	Subtraction Patterns										
		S1	S2	S3	S4	S5	S6	S7	S8	S9	S10	Total
Grade 3	1-step	51	15	2	22	37	9	5	2	0	14	157
	Hidden-step	0	0	0	0	0	0	0	0	0	0	0
	Final Step	1	1	0	0	0	1	0	0	0	0	3
	Total	52	16	2	22	37	10	5	2	0	14	160
Grade 4	1-step	27	18	3	15	52	15	1	19	0	12	162
	Hidden-step	5	0	1	0	0	0	0	1	0	1	8
	Final Step	4	6	0	0	2	0	0	2	0	0	14
	Total	36	24	4	15	54	15	1	22	0	13	184
Grade 5	1-step	23	7	2	8	63	30	0	28	1	11	173
	Hidden-step	14	1	1	1	0	0	0	13	1	2	33
	Final Step	14	12	0	5	9	2	1	3	0	1	47
	Total	51	20	3	14	72	32	1	44	2	14	253
Grade 6	1-step	10	1	1	1	38	20	6	25	0	15	117
	Hidden-step	5	0	2	0	4	0	0	8	0	2	21
	Final Step	28	6	1	4	13	3	0	1	0	5	61
	Total	43	7	4	5	55	23	6	34	0	22	199

Summary for Grades 3 to 6 Inclusive

	S1	S2	S3	S4	S5	S6	S7	S8	S9	S10	Total
Total	182	67	13	56	218	80	13	102	2	63	796
%	22.9	8.4	1.6	7.0	27.4	10.1	1.6	12.8	0.3	7.9	100
Rank	2	5	8	7	1	4	8	3	10	6	

TABLE XIVc --- TEXT Z

Summary of Subtraction Patterns Found in the One-Step and Multiple-Step Verbal Problems of Text Z

	Type of Repetition	Subtraction Patterns										
		S1	S2	S3	S4	S5	S6	S7	S8	S9	S10	Total
Grade 3	1-step	59	18	21	43	28	18	0	25	1	15	228
	Hidden-step	1	0	0	0	1	0	0	0	0	0	2
	Final Step	1	1	0	0	0	0	0	0	0	0	2
	Total	61	19	21	43	29	18	0	25	1	15	232
Grade 4	1-step	34	7	6	19	18	5	1	10	2	4	106
	Hidden-step	0	0	1	1	1	0	0	0	0	0	3
	Final Step	7	13	0	3	0	0	0	8	0	0	31
	Total	41	20	7	23	19	5	1	18	2	4	140
Grade 5	1-step	34	4	3	9	19	6	5	8	2	6	96
	Hidden-step	6	0	0	3	1	2	0	11	0	0	23
	Final Step	20	4	0	0	4	2	0	33	0	0	63
	Total	60	8	3	12	24	10	5	52	2	6	182
Grade 6	1-step	19	2	6	5	7	7	6	20	0	3	75
	Hidden-step	1	0	3	0	1	0	0	8	0	0	13
	Final Step	16	2	1	1	3	1	1	15	0	2	42
	Total	36	4	10	6	11	8	7	43	0	5	130

Summary for Grades 3 to 6 Inclusive

	S1	S2	S3	S4	S5	S6	S7	S8	S9	S10	Total
Total	198	51	41	84	83	41	13	138	5	30	684
%	28.9	7.5	6.0	12.3	12.1	6.0	1.9	20.2	0.7	4.4	100
Rank	1	5	6	3	4	6	9	2	10	8	

TABLE XIVd --- TEXT W

Summary of Subtraction Patterns Found in the One-Step and Multiple-Step Verbal Problems of Text W

	Type of Repetition	Subtraction Patterns										
		S1	S2	S3	S4	S5	S6	S7	S8	S9	S10	Total
Grade 3	1-step	70	10	12	25	31	11	3	3	0	15	180
	Hidden-step	0	0	0	0	0	0	0	0	0	0	0
	Final Step	0	0	0	0	0	0	0	0	0	0	0
	Total	70	10	12	25	31	11	3	3	0	15	180
Grade 4	1-step	32	24	7	29	15	6	2	3	0	4	122
	Hidden-step	0	0	0	0	0	0	0	0	0	0	0
	Final Step	11	14	0	9	4	0	0	3	0	0	41
	Total	43	38	7	38	19	6	2	6	0	4	163
Grade 5	1-step	41	3	2	20	18	19	8	12	4	7	134
	Hidden-step	1	0	0	1	1	0	0	4	0	0	7
	Final Step	28	37	2	15	4	1	0	8	0	2	97
	Total	70	40	4	36	23	20	8	24	4	9	238
Grade 6	1-step	14	2	1	5	12	22	9	18	1	11	95
	Hidden-step	6	0	0	0	3	0	0	12	0	2	23
	Final Step	22	2	0	4	3	0	1	9	0	4	45
	Total	42	4	1	9	18	22	10	39	1	17	163

Summary for Grades 3 to 6 Inclusive

	S1	S2	S3	S4	S5	S6	S7	S8	S9	S10	Total
Total	225	92	24	108	91	59	23	72	5	45	744
%	30.2	12.4	3.2	14.5	12.2	7.9	3.1	9.7	0.7	6.1	100
Rank	1	3	8	2	4	6	9	5	10	7	

TABLE XVa --- TEXT X

Summary of Multiplication Patterns Found in the One-Step and Multiple-Step Verbal Problems of Text X

	Type of Repetition	Multiplication Patterns								
		M1	M2	M3	M4	M5	M6	M7	M8	Total
Grade 3	1-step	58	89	4	0	0	0	0	1	152
	Hidden-step	0	5	0	0	0	0	0	0	5
	Final Step	0	0	0	0	0	0	0	0	0
	Total	58	94	4	0	0	0	0	1	157
Grade 4	1-step	62	83	8	0	0	0	0	0	153
	Hidden-step	17	58	0	0	0	0	0	0	75
	Final Step	5	14	0	0	0	0	0	0	19
	Total	84	155	8	0	0	0	0	0	247
Grade 5	1-step	34	61	9	1	8	8	0	0	121
	Hidden-step	18	68	0	0	19	3	0	0	108
	Final Step	4	28	0	0	3	1	0	0	36
	Total	56	157	9	1	30	12	0	0	265
Grade 6	1-step	32	32	9	3	16	9	39	3	143
	Hidden-step	8	32	2	0	6	4	30	0	82
	Final Step	9	33	2	0	5	2	2	0	53
	Total	49	97	13	3	27	15	71	3	278

Summary for Grades 3 to 6 Inclusive

	M1	M2	M3	M4	M5	M6	M7	M8	Total
Total	247	503	34	4	57	27	71	4	947
%	26.2	53.1	3.6	0.4	6.0	2.8	7.5	0.4	100
Rank	2	1	5	7	4	6	3	7	

TABLE XVb --- TEXT Y

Summary of Multiplication Patterns Found in the One-Step and Multiple-Step Verbal Problems of Text Y

	Type of Repetition	Multiplication Patterns								
		M1	M2	M3	M4	M5	M6	M7	M8	Total
Grade 3	1-step	62	70	3	0	0	0	0	0	135
	Hidden-step	0	3	0	0	0	0	0	0	3
	Final Step	0	0	0	0	0	0	0	0	0
	Total	62	73	3	0	0	0	0	0	138
Grade 4	1-step	77	117	2	1	0	0	0	1	198
	Hidden-step	4	15	0	0	0	0	0	0	19
	Final Step	3	15	1	0	0	0	0	0	19
	Total	84	147	3	1	0	0	0	1	236
Grade 5	1-step	55	92	3	1	12	9	0	1	173
	Hidden-step	13	60	0	0	9	0	0	0	82
	Final Step	8	42	0	0	3	7	0	0	60
	Total	76	194	3	1	24	16	0	1	315
Grade 6	1-step	44	66	4	1	11	3	21	11	161
	Hidden-step	17	34	0	0	22	2	29	0	104
	Final Step	12	74	0	0	12	14	4	0	116
	Total	73	174	4	1	45	19	54	11	381

Summary for Grades 3 to 6 Inclusive

	M1	M2	M3	M4	M5	M6	M7	M8	Total
Total	295	588	13	3	69	35	54	13	1070
%	27.6	55.0	1.2	0.3	6.4	3.3	5.0	1.2	100
Rank	2	1	6	8	3	5	4	6	

TABLE XVc --- TEXT Z

Summary of Multiplication Patterns Found in the One-Step and Multiple-Step Verbal Problems of Text Z

	Type of Repetition	Multiplication Patterns								
		M1	M2	M3	M4	M5	M6	M7	M8	Total
Grade 3	1-step Hidden-step Final Step	75 1 0	73 9 4	0 0 0	0 0 0	0 0 0	0 0 0	0 0 0	0 0 0	148 10 4
	Total	76	86	0	0	0	0	0	0	162
Grade 4	1-step Hidden-step Final Step	68 13 7	126 41 35	7 0 0	0 0 0	0 0 0	0 0 0	0 0 0	0 0 0	201 54 42
	Total	88	202	7	0	0	0	0	0	297
Grade 5	1-step Hidden-step Final Step	35 35 11	64 69 48	1 0 0	0 0 0	18 23 6	14 4 0	0 0 0	1 0 0	133 131 65
	Total	81	181	1	0	47	18	0	1	329
Grade 6	1-step Hidden-step Final Step	30 17 18	72 26 61	12 0 0	1 0 0	23 8 16	12 2 8	14 5 0	1 4 3	165 62 106
	Total	65	159	12	1	47	22	19	8	333

Summary for Grades 3 to 6 Inclusive

	M1	M2	M3	M4	M5	M6	M7	M8	Total
Total % Rank	310 27.7 2	628 56.0 1	20 1.8 5	1 0.1 8	94 8.4 3	40 3.5 4	19 1.7 6	9 0.8 7	1121 100

TABLE XVd --- TEXT W

Summary of Multiplication Patterns Found in the One-Step and Multiple-Step Verbal Problems of Text W

	Type of Repetition	Multiplication Patterns								
		M1	M2	M3	M4	M5	M6	M7	M8	Total
Grade 3	1-step	41	74	0	0	0	0	0	10	125
	Hidden-step	0	0	0	0	0	0	0	0	0
	Final Step	0	0	0	0	0	0	0	0	0
	Total	41	74	0	0	0	0	0	10	125
Grade 4	1-step	116	140	11	0	0	0	0	0	267
	Hidden-step	10	43	1	0	0	0	0	0	54
	Final Step	1	0	0	0	0	0	0	0	1
	Total	127	183	12	0	0	0	0	0	322
Grade 5	1-step	58	56	10	0	6	15	0	0	145
	Hidden-step	26	81	0	0	10	0	0	2	119
	Final Step	7	45	0	0	13	0	0	0	65
	Total	91	182	10	0	29	15	0	2	329
Grade 6	1-step	52	58	9	1	9	19	27	4	179
	Hidden-step	26	32	0	0	19	2	27	2	108
	Final Step	13	95	0	0	17	29	4	0	158
	Total	91	185	9	1	45	50	58	6	445

Summary for Grades 3 to 6 Inclusive

	M1	M2	M3	M4	M5	M6	M7	M8	Total
Total	350	624	31	1	74	65	58	18	1221
%	28.7	51.1	2.5	0.1	6.1	5.3	4.7	1.5	100
Rank	2	1	6	8	3	4	5	7	

TABLE XVIa --- TEXT X

Summary of Division Patterns Found in the One-Step and Multiple-Step Verbal Problems of Text X

	Type of Repetition	Division Patterns																
		D1	D2	D3	D4	D5	D6	D7	D8	D9	D10	D11	D12	D13	D14	D15	D16	Total
Grade 3	1-step	20	36	8	0	0	0	0	0	34	2	19	6	0	0	0	1	126
	Hidden-step	0	0	0	0	0	0	0	0	0	0	0	0	0	0	0	0	0
	Final Step	0	0	0	0	0	0	0	0	0	0	0	0	0	0	0	0	0
	Total	20	36	8	0	0	0	0	0	34	2	19	6	0	0	0	1	126
Grade 4	1-step	60	35	2	1	0	0	7	2	24	21	34	14	4	15	0	0	219
	Hidden-step	1	1	0	0	0	0	0	0	6	0	1	0	0	0	0	0	9
	Final Step	2	1	0	0	0	0	2	1	1	3	12	4	27	6	0	0	59
	Total	63	37	2	1	0	0	9	3	31	24	47	18	31	21	0	0	287
Grade 5	1-step	30	11	4	7	0	0	13	7	0	5	28	16	19	30	7	0	177
	Hidden-step	6	0	0	1	0	0	0	0	0	0	0	12	0	0	0	0	19
	Final Step	5	1	2	0	0	0	0	0	0	1	12	1	26	6	0	0	54
	Total	41	12	6	8	0	0	13	7	0	6	40	29	45	36	7	0	250
Grade 6	1-step	19	2	2	7	2	1	18	23	0	16	7	14	14	17	3	0	145
	Hidden-step	3	0	0	1	0	0	15	3	0	0	2	0	0	0	0	0	24
	Final Step	3	0	1	0	0	0	1	2	0	1	5	0	16	5	1	0	35
	Total	25	2	3	8	2	1	34	28	0	17	14	14	30	22	4	0	204

Summary for Grades 3 to 6 Inclusive

	D1	D2	D3	D4	D5	D6	D7	D8	D9	D10	D11	D12	D13	D14	D15	D16	Total
Total	149	87	19	17	2	1	56	38	65	49	120	67	106	79	11	1	867
%	17.2	10.0	2.2	2.0	0.2	0.1	6.5	4.4	7.5	5.7	13.8	7.7	12.2	9.1	1.3	0.1	100
Rank	1	4	11	12	14	15	8	10	7	9	2	6	3	5	13	15	

TABLE XVIb --- TEXT Y

Summary of Division Patterns Found in the One-Step and Multiple-Step Verbal Problems of Text Y

	Type of Repetition	Division Patterns																
		D1	D2	D3	D4	D5	D6	D7	D8	D9	D10	D11	D12	D13	D14	D15	D16	Total
Grade 3	1-step	18	17	1	0	0	0	0	0	20	24	5	13	1	0	0	0	99
	Hidden-step	0	0	0	0	0	0	0	0	0	0	0	0	0	0	0	0	0
	Final Step	0	0	0	0	0	0	0	0	0	0	0	0	0	0	0	0	0
	Total	18	17	1	0	0	0	0	0	20	24	5	13	1	0	0	0	99
Grade 4	1-step	30	37	2	0	1	0	3	1	20	23	14	14	14	4	0	1	164
	Hidden-step	0	0	0	0	0	0	0	0	7	1	1	3	0	1	0	0	13
	Final Step	1	3	0	0	0	0	0	0	0	3	2	0	6	1	0	0	16
	Total	31	40	2	0	1	0	3	1	27	27	17	17	20	6	0	1	193
Grade 5	1-step	35	19	7	9	1	0	1	2	0	10	6	27	34	11	1	1	164
	Hidden-step	10	0	0	0	0	0	0	14	0	1	3	13	0	0	0	0	41
	Final Step	4	7	3	0	0	0	0	2	0	4	1	5	10	1	0	1	38
	Total	49	26	10	9	1	0	1	18	0	15	10	45	44	12	1	2	243
Grade 6	1-step	42	3	7	5	1	1	20	18	0	13	0	22	20	14	2	0	168
	Hidden-step	20	1	0	2	0	0	0	33	0	3	0	16	2	0	0	0	77
	Final Step	3	1	1	0	0	0	0	4	0	0	1	1	11	0	2	0	24
	Total	65	5	8	7	1	1	20	55	0	16	1	39	33	14	4	0	269

Summary for Grades 3 to 6 Inclusive

	D1	D2	D3	D4	D5	D6	D7	D8	D9	D10	D11	D12	D13	D14	D15	D16	Total
Total	163	88	21	16	3	1	24	74	47	82	33	114	98	32	5	3	804
%	20.3	10.9	2.6	2.0	0.4	0.1	3.0	9.2	5.8	10.2	4.1	14.2	12.2	4.0	0.6	0.4	100
Rank	1	4	11	12	14	16	10	6	7	5	8	2	3	9	13	14	

TABLE XVIc --- TEXT Z

Summary of Division Patterns Found in the One-Step and Multiple-Step Verbal Problems of Text Z

	Type of Repetition	Division Patterns																
		D1	D2	D3	D4	D5	D6	D7	D8	D9	D10	D11	D12	D13	D14	D15	D16	Total
Grade 3	1-step	36	42	2	0	0	0	1	0	31	7	26	9	0	0	0	0	154
	Hidden-step	1	0	0	0	0	0	0	0	2	0	0	0	0	0	0	0	3
	Final Step	0	0	0	0	0	0	0	0	0	0	0	0	0	0	0	0	0
	Total	37	42	2	0	0	0	1	0	33	7	26	9	0	0	0	0	157
Grade 4	1-step	43	48	3	9	0	1	1	0	18	24	66	27	5	7	1	0	253
	Hidden-step	6	0	0	0	0	0	15	1	28	0	0	4	0	0	0	0	54
	Final Step	2	0	1	0	0	0	0	0	1	0	2	0	15	0	0	0	21
	Total	51	48	4	9	0	1	16	1	47	24	63	31	20	7	1	0	328
Grade 5	1-step	51	41	8	19	1	1	1	1	0	14	24	31	7	10	2	0	211
	Hidden-step	36	0	0	0	1	0	11	4	0	0	1	11	0	0	0	0	64
	Final Step	8	7	5	0	0	0	0	0	0	0	0	10	11	0	0	0	41
	Total	95	48	13	19	2	1	12	5	0	14	25	52	18	10	2	0	316
Grade 6	1-step	62	4	9	5	2	1	3	7	0	8	12	16	21	7	2	0	159
	Hidden-step	26	0	0	1	0	0	35	32	0	2	0	3	1	2	0	0	102
	Final Step	5	1	1	0	1	0	1	1	0	0	1	1	5	1	3	0	21
	Total	93	5	10	6	3	1	39	40	0	10	13	20	27	10	5	0	282

Summary for Grades 3 to 6 Inclusive

	D1	D2	D3	D4	D5	D6	D7	D8	D9	D10	D11	D12	D13	D14	D15	D16	Total
Total	276	143	29	34	5	3	68	46	80	55	132	112	65	27	3	0	1083
%	25.5	13.2	2.7	3.1	0.5	0.3	6.3	4.2	7.4	5.1	12.2	10.3	6.0	2.5	0.7	0.0	100
Rank	1	2	11	10	14	15	6	9	5	8	3	4	7	12	13	16	

TABLE XVId --- TEXT W

Summary of Division Patterns Found in the One-Step and Multiple-Step Verbal Problems of Text W

	Type of Repetition	Division Patterns																
		D1	D2	D3	D4	D5	D6	D7	D8	D9	D10	D11	D12	D13	D14	D15	D16	Total
Grade 3	1-step	18	36	5	0	0	0	0	0	34	27	9	10	0	0	0	1	140
	Hidden-step	0	0	0	0	0	0	0	0	0	0	0	0	0	0	0	0	0
	Final Step	0	0	0	0	0	0	0	0	0	0	0	0	0	0	0	0	0
	Total	18	36	5	0	0	0	0	0	34	27	9	10	0	0	0	1	140
Grade 4	1-step	51	31	3	1	4	0	2	0	23	33	26	48	11	7	0	1	246
	Hidden-step	0	0	0	0	0	0	0	0	18	0	0	0	0	0	0	0	18
	Final Step	0	0	0	0	0	0	0	0	0	2	0	0	6	0	0	0	8
	Total	51	31	8	1	4	0	2	0	41	35	26	48	17	7	0	1	272
Grade 5	1-step	24	12	0	5	0	0	1	6	0	36	17	26	30	25	2	5	189
	Hidden-step	12	0	0	0	0	0	20	13	0	0	0	2	0	0	0	0	47
	Final Step	1	2	0	0	0	0	0	0	0	0	1	1	19	1	0	0	25
	Total	37	14	0	5	0	0	21	19	0	36	18	29	49	26	2	5	261
Grade 6	1-step	33	11	7	0	0	0	2	14	0	14	5	19	10	10	0	0	125
	Hidden-step	27	0	0	0	0	0	28	28	0	0	0	4	0	0	0	1	88
	Final Step	8	0	3	0	0	0	1	6	0	2	0	1	10	1	0	1	33
	Total	68	11	10	0	0	0	31	48	0	16	5	24	20	11	0	2	246

Summary for Grades 3 to 6 **Inclusive**

	D1	D2	D3	D4	D5	D6	D7	D8	D9	D10	D11	D12	D13	D14	D15	D16	Total
Total	174	92	23	6	4	0	54	67	75	114	58	111	86	44	2	9	919
%	18.9	10.0	2.5	0.6	0.4	0.0	5.9	7.3	8.2	12.4	6.3	12.1	9.4	4.8	0.2	1.0	100
Rank	1	4	11	13	14	16	9	7	6	2	8	3	5	10	15	12	